NEW TOP COMMERCIAL SPACE Ⅱ

中国最新顶尖商业空间Ⅱ

深圳市创扬文化传播有限公司 编
赵 欣 译

大连理工大学出版社

图书在版编目(CIP)数据

中国最新顶尖商业空间. 2：汉英对照／深圳市创扬文化传播有限公司编；赵欣译.—大连：大连理工大学出版社，2009.5

ISBN 978-7-5611-4777-1

Ⅰ.中… Ⅱ.①深…②赵… Ⅲ.商业—服务建筑—室内设计：空间设计—中国—汉、英 Ⅳ.TU247

中国版本图书馆CIP数据核字（2009）第049799号

出版发行：大连理工大学出版社

（地址：大连市软件园路80号　邮编：116023）

印　　刷：利丰雅高印刷（深圳）有限公司

幅面尺寸：242mm × 280mm

印　　张：17

插　　页：4

出版时间：2009年5月第1版

印刷时间：2009年5月第1次印刷

策　　划：袁　斌

责任编辑：刘　蓉

责任校对：雷　娜

封面设计：温广强

版式设计：温广强

ISBN 978-7-5611-4777-1

定　　价：228.00元

电 话：0411-84708842

传 真：0411-84701466

邮 购：0411-84703636

E-mail：designbooks_dutp@yahoo.cn

URL：http://www.dutp.com.cn

如有质量问题请联系出版中心：（0411）84709246　84709043

Preface
前言

Keiichiro Sako
SAKO Architects
迫庆一郎
北京卅口建筑设计咨询有限公司

Create Non-Daily Space

Nowadays, in the intense competition of business, in order to attract more customers, to make the shop more attractive, to set off the charm of goods, and to provide users with better space for activities, commercial space design has become an indispensable strategy by many merchants to promote their products. Therefore, to a designer how to create an attractive commercial space has become an unavoidable issue when he does commercial design. In my opinion, the main feature and charm of commercial design lie in its activity to create non-daily space. From the viewpoint of consumers, the idea of consumption appears perhaps due to the pursuit of good material life, perhaps due to the boredom of common life. However, it is actually not an easy thing to find a suitable item quickly and accurately in the vast sea of products. Precisely because of this, many merchants hope to attract more business opportunities by shop space which is a "media" immediately stimulating consumer senses. In the shop space, the designers repackage the products through space, and the main component of "packaging material" is composed of design theme. The theme, just like a bridge, links the products with the corresponding end-users. Put it another way, if the design of commercial space is compared to a "theme park", it will create an area divided from the outside world, at the same time lead us to another space which is full of imagination. However, it is different from the theme park, where the theme of commercial space is neither to reproduce any street in the real world, nor to show the existing characters as prototypes in the new world, but it fully absorbs nutrients from daily life, and resets and re-edits them, so that the new space is endowed with different elements from daily life. Therefore, the design theme of commercial space is not borrowed from anywhere, but it is a new generation based on the daily life.

打造非日常空间

在商业竞争激烈的今天，为了能争取更多的客户源，为了使商铺更加有吸引力，并更能衬托出商品的魅力以及给用户提供更加良好的活动空间，商业空间设计成了许多商家运用在产品推销上的一个不可缺少的策略。因此，对于一名设计师来说，如何能打造出一个具有吸引力的商业空间成了在做商业设计时一个不可避免的问题。

我认为商业设计的主要特色与魅力在于它打造非日常性空间的行为。从消费者的角度出发，也许是因为对良好物质生活的追求，也许是因为对平凡生活的厌倦，才会出现消费的念头。但是在浩瀚的产品海洋里能够快速准确地找到适合自己的物品其实并不是件容易的事情。也正因为如此，许多商家都希望利用商铺空间这样一种可立即刺激消费者感观的"媒体"去吸引更多的商业机会。在这里，设计师通过空间将产品重新包装。而"包装材料"的主要成分由设计主题组成。在这里，主题就好比一座衔接产品与其相应终端用户的桥梁。换一种说法，如果把商业空间设计比做"主题公园"，那么它营造出一个被外界隔离的领域，同时引领人们进入另一个充满遐想的空间。然而与主题公园不同的是，商业空间的主题并不是重现现实世界中的某个街道，也不是把已有人物作为原型在新的世界里展开，而是从日常生活中充分吸取养分，并将其重新组合编辑，从而使新的空间里有着与日常生活不同的要素。因此，商业空间设计中的主题不是借来的，而是基于日常生活新产生出来的。

Hongbin Pan
PANORAMA International Ltd.
潘鸿彬
泛纳国际设计顾问有限公司

The interior design has entered our lives, in a wide range of design industries, the interior design has become an integral part of daily life, and we can feel our own environment through the interior design.
Nowadays, the interior design has received more and more attention, and has been accepted by citizens. It provides prospective pictures for the modern society, and integrates the commercial elements and cultural characteristics. The burden on contemporary designers is hard to bear, and we are working hard to create better living and working environment for people.
A large number of facts have proven that the design industry has become an element of international competition, and many advanced countries demonstrate their designers' creative genius to try to make themselves to be the international design center.
Now I would like to work together with our designers, and with the pace of our economic development, we will develop our design quickly to the contemporary level of international standards.

室内设计已步入了我们的生活，在各式各样的设计行业中，室内设计已成为人们日常生活的组成部分，通过室内设计，我们可以感受到自己所处的环境。

现在室内设计这一行已越来越受到重视，并为人们所接受。它给现代社会提供美好的图景，将商业元素和文化特质融合在一起。当代设计人才肩负重任，大家都在努力为人类创造更美好的生活和工作环境。

大量事实证明，设计这个行业已成为国际竞争的一个元素，许多先进国家均纷纷展示其设计师的创作天才，力图将自己变成国际设计中心。

在此，本人希望和大家一道共同努力，随着我国经济发展的步伐，迅速将我们的设计水平提高到当代国际水准。

Contents
目录

Gallery More	Gallery More	006
Anicée Showroom Design	Anicée Showroom Design	018
Dashing Diva	Dashing Diva	026
DJS	DJS缀饰	034
Chinese Arts & Crafts	中艺品牌店	040
YXZD Clothing Shop	“一席之地”服饰店	048
Stage for Hair	发舞台	054
ROMANTICISM 2 in Hangzhou	杭州浪漫一身2	062
EIFINI in Chengdu	成都伊芙丽	070
KID'S REPUBLIC in Shanghai	上海蒲蒲兰绘本馆	078
BRANCH in Changchun	长春藤	090
FLATFLAT in Harajuku	原宿FLATFLAT	100
IMAGINARIUM in Barcelona	巴塞罗那IMAGINARIUM	108
GOBI in Ulaanbaatar	乌兰巴托GOBI	114
PAMPER HEiRESS	PAMPER HEiRESS	118
Giant Bicycle Inc.'s Flagship Store	捷安特女性旗舰店	126
FZL Concept Shop	FZL概念店	136
Bristol Hair Salon	布里斯托美发沙龙	146

Yogo Juice in Jinmao, Shanghai	Yogo Juice上海金茂店	152
J.K. Salon	J.K.Salon	156
T.Y.M(International)Hair Salon	T.Y.M(国际)发型	162
Bowling Gym	保龄球馆	168
Fuzhou Dynamic Beauty Salon	福州动感形象沙龙	176
Futu Bridal Shop	富图婚纱店	182
Gama Photography Gallery	伽玛影研	190
YOYI Book	深圳友谊书城	196
Iridescence o0O	虹色o0O	202
List of Popular Jewelry	珠宝流行榜	210
B&F Cosmetics Franchised Store in Zhongxin Square	中信广场爱敬化妆品专卖店	218
Japanese WATABE (Shanghai) Wedding Studio	日本华德培(上海)婚庆工作室	224
Pure Pearl Store	千足珍珠示范店	230
Croissants de France in Vanke Town	可颂坊万科城店	236
Mingcheng Tea Supermarket	茗城茶业超市	246
Dongpeng Sanitary Appliance	东鹏洁具	254
Shandong Jinluo Group	山东金锣集团	262

Gallery More

Gallery More

地点/中国台北　面积/270平方米　设计师/沈志忠
设计单位/建构线设计有限公司
主要材料/不锈钢毛丝面、黑板岩、黑色烤漆玻璃、钢丝线、柚木、玻璃、塑铝板、黑色大理石

This project provides us another alternative of space experience in a clothing shop; this concept introduces experience of visiting a gallery into the design of a clothing shop, treats the clothes as works of art, and brings the experience of visiting, admiring and enjoying in the gallery into this space. The most important element is to design a wall with more than ten thousand steel cables, at the same time, these cables are used to support those iron frameworks. The cables represent the most basic material of clothes—braided wire, so the cables reflect the raw material of clothes. The clothes are displayed in the framework, like priceless paintings, and the deliberately-designed distance is used to show the high quality and workmanship of the clothes. The cables, the iron-gray mirror behind the glass, walls, open space and exquisite lighting are integrated to create a fashionable space with different feeling, like an art gallery.

GALLERY MORE

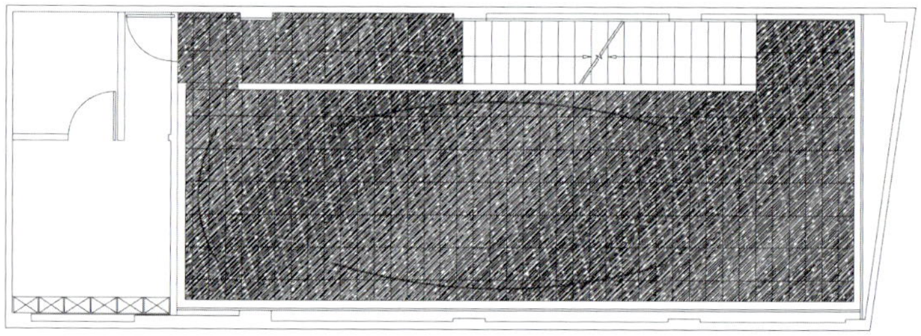

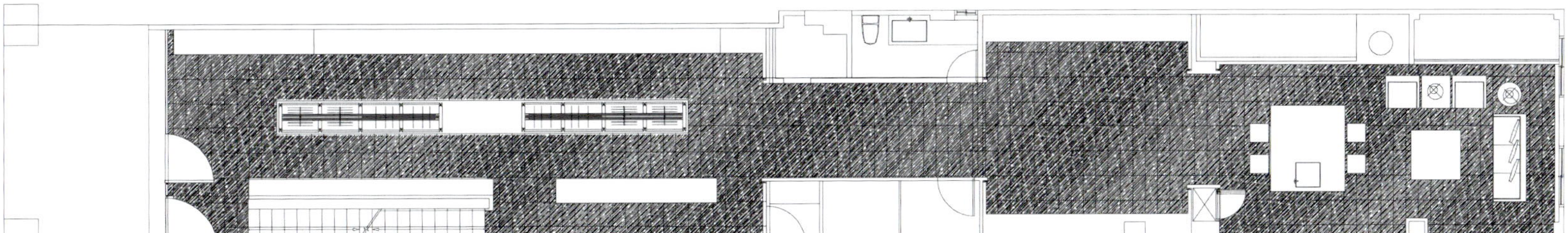

本案提供了一种另类逛服饰店的空间经验，这个概念是把逛美术馆的经验带进服饰店的设计之内，将服饰当做艺术品来看待，将在美术馆观看与欣赏、感受的经验行为带进空间。在本案中最主要的一个元素就是用超过一万条的钢索去转换成墙面，这些钢索也同时用以支撑那些铁质的框架，钢索代表的是衣服最基本的材料——织线，所以，借着钢索反映衣服的原料。衣服被框在框架里，像无价的画作，藉由被刻意拉开的距离，去呈现服装的高质量与工法。钢索、玻璃后面的铁灰的镜面、墙面、开放的空间、精致的灯光，整合起来创造出一个像艺廊似的，拥有不一样感受的时尚空间。

Anicée Showroom Design

Anicée Showroom Design

地点／中国台北　面积／约86平方米　设计师／戴源昌
设计单位／达阳室内装修有限公司
主要材料／铁管、锈铁、玻璃、epoxy、木作、喷漆
摄影师／范宸雄

This is a clothing shop with a self-created brand. This design applies the concept of environmental protection which is popular at present, and the space is designed without excessive decorating and packaging. The simple and clean grayish white is embellished with pole accessories of free curves, and the clothes are the only protagonists in this space. The show space on the first floor is not large, but the glasses of large area without frames on the three facades extend the internal and external spaces, so there seems to be an imperceptible dialogue between the passers-by and the clothes. The clean facades and the unsmooth floor covered with seamless Epoxy make the space natural, simple and strong. Pieces of clothes are hanging on the pole accessories, as if there is breeze swaying them.

The original circular staircase extends the center column up continuously, and the extension to left and right seems to be showing its vitality. The same light gray Epoxy floor and the original walls which are re-whitened enlarge the space on vision. Besides the pole accessories poping their heads, there are white laminates twisting secretly from the wall. The three large floor doors of mirrors seem to be discretional, which extend the space to other place, faraway and broad.

anicée
anicée

anicée

这是一家自创品牌的服装店。设计采用时下流行的环保概念，空间无过度的装饰与装潢，简单干净的灰白色调加上自由曲线的杆件，空间内惟一的主角就是商品。一楼展示空间不大，三向外观立面用无框边的大片玻璃内外延伸，无形中路人与商品似乎有了对话。干净的立面与不平整的地面覆盖无缝Epoxy，让空间看起来自然且简洁有力。杆件上垂挂着一件件的衣服，彷佛有一阵阵微风向它们吹拂着。

原有的圆形梯将中心立柱持续向上延伸，左右的伸展，似乎是在表现它的生命力，相同的浅灰色Epoxy地面与原有墙面重新刷白，让空间在视觉上放大许多。这里除了有一支支探头出来的杆件外，还多了从壁面偷偷扭曲出来的白色层板。三块大片的落地镜门像是任意门，把空间拉向既遥远又宽广的另一方。

anicede

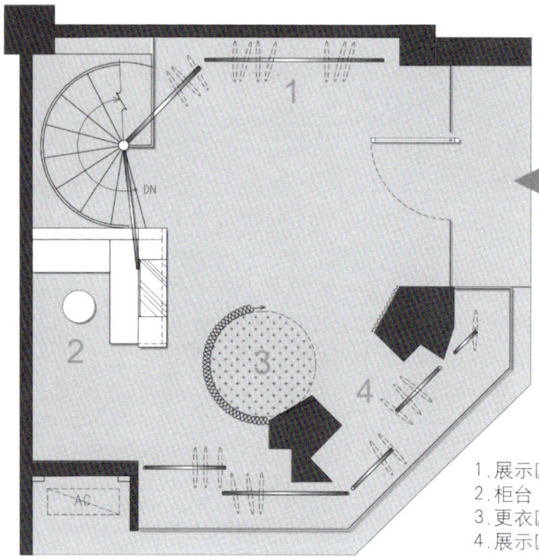
1
2
3
4
DN
AC
1.展示区
2.柜台
3.更衣区
4.展示区

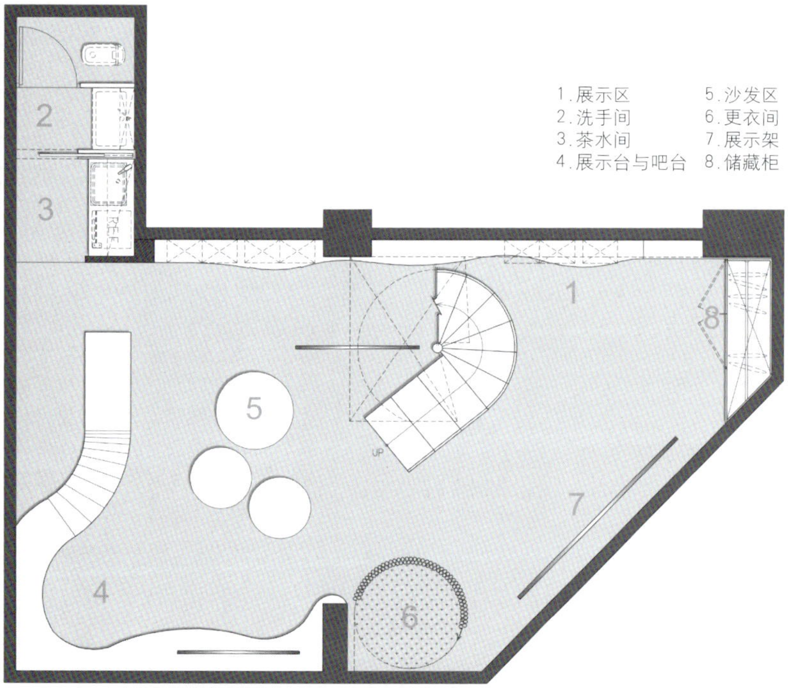
1.展示区
2.洗手间
3.茶水间
4.展示台与吧台
5.沙发区
6.更衣间
7.展示架
8.储藏柜
1
2
3
4
5
6
7
8
UP

Dashing Diva

Dashing Diva

The designers hope to use the arc to reflect the gentle and natural free lines of nails. Pink is the identity color of the enterprise, Dashing Diva, that naturally becomes the main color of the whole shop. Overall, designers would like to express the beauty of sex and sentiment. Designers try to use the Retro approach by using a big curved shape roof for the Pedicure Area, and with the lava lamps bubbles that presents a lovely and relaxing environment. For the retail display area, designers apply the concept of curve to design the display counters. For the nail products, designers just simply use a transparent acrylic to support the entire wall covered with products to emphasize its richness. For the ceiling of the second floor, designers' expression of design becomes more flexible. The designers employ the Optical Arts as the wall background, so that they can place products, which will have a particular three-dimensional vision of goods.

面积/350平方米　设计师/赵牧桓
参与设计/方凯、罗玉龙、陈君正、义明辉
设计单位/MoHen Design International
主要材料/人造石、塑胶地砖、铁件、亚克力、密度板
摄影师/周宇贤

Nail Extension
Acrylic Nails
UV Gel Nails
Soak Off And Removal
Tailor Fit Application

DASHING
DIVA
DASHING
DIVA

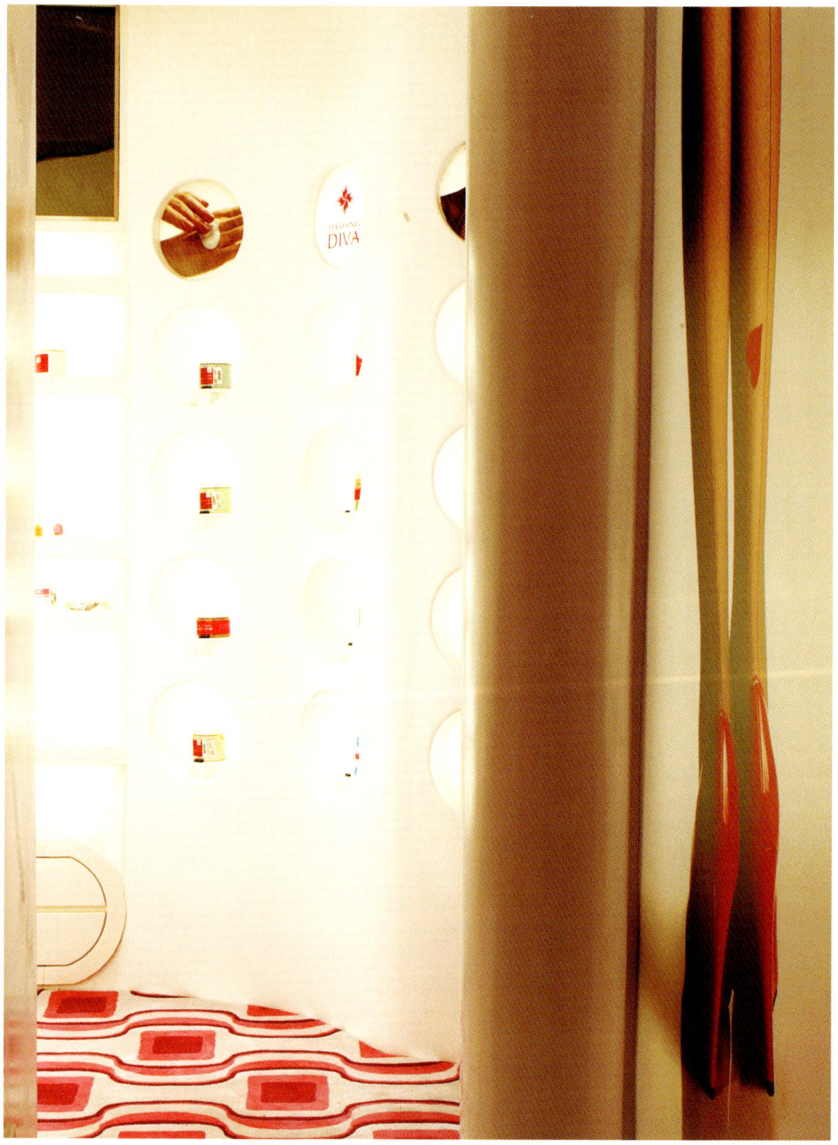

DIVA

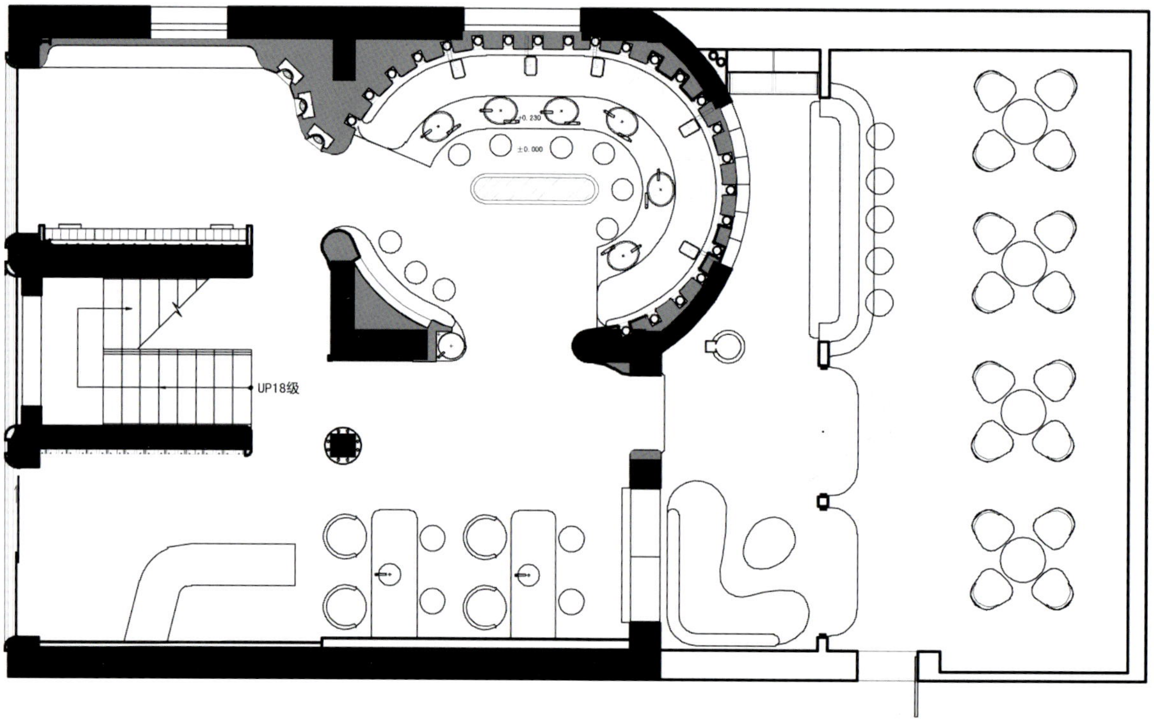

+0. 230
±0. 000
UP18级

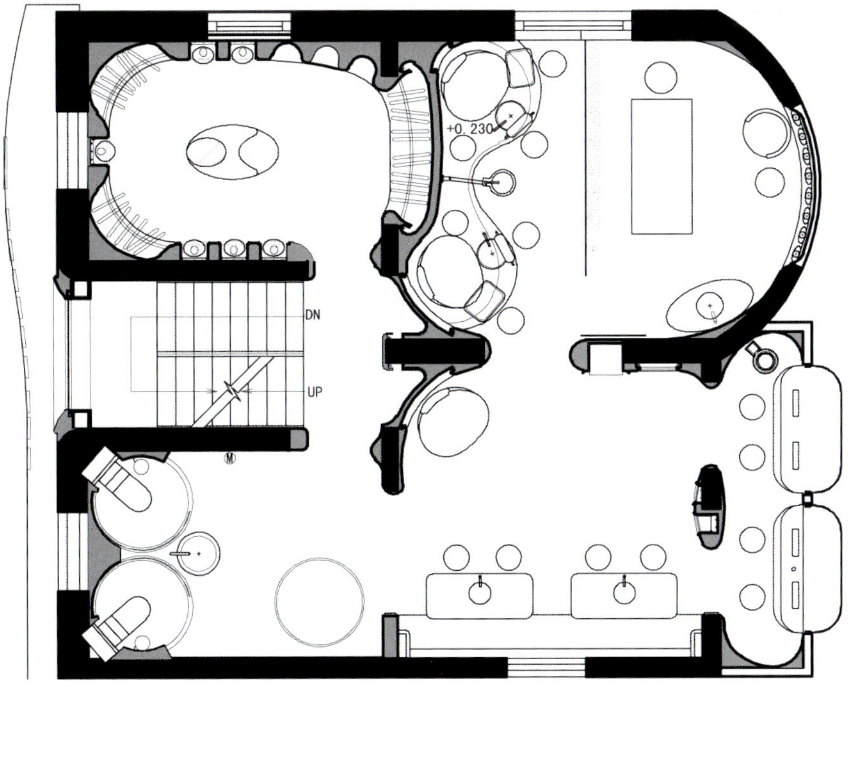

设计师尽量用弧形来表现指甲温柔自由的天生线条。粉红色是Dashing Diva主要的企业识别色调，很自然地就成为了本案的色彩基调。设计师希望表达一种性感和感性的美丽。在美脚区，设计师运用复古的手法设计了一个大的弧形棚盖，并用熔岩灯所衍生的泡泡强调了它的慵懒和可爱。在零售展示区，设计师运用弧形的概念来设计展示柜。在美甲产品展示区，设计师用透明亚克力把产品支撑起来，布满整面墙，强调它的丰富性。二楼的天花设计，设计师的表现手法就更为自由，设计师还把背景墙设计成立体构成的形式，让店面陈列摆设自己的商品，强调商品的立体感和特殊性。

PLEASE GO UPSTAIRS

GO UPSTAIRS

DJS

DJS缀饰

地点/中国香港　面积/150平方米　设计师/潘鸿彬
设计单位/泛纳国际设计顾问有限公司
主要材料/云石、镜钢、背漆玻璃、发光石、人造皮

The Beauty of Facets

The interior designing strategy of DJS is to make use of the various facets, along with the common natures and chemical structure of diamonds and jades to generate a unique identity for the brand. To achieve this, the interior design makes use of the metaphor of interpreting the retail space as a glowing gem for the customer to explore. The white spatial envelope of glowing alabaster showcase wall units and white marble floor, are firstly set up to provide a minimal setting for the space. The symmetrical layout is then adopted to echo the squareness of the site. This is followed by combining the geometrical composition of facets. Such a design is of both aesthetic and functional effects. White glowing wall units with tilted vertical faces in random angles allow green side facings to be exposed. This subtle addition of green depicts the natural beauty of jade's colour and translucency. Rows of floating glazed "showcase squares" with different sizes and heights create rhythm and strong visual attractions for the customers. In these "showcase squares", there are custom-made frosted acrylic display stands. The interplay of angular language is finally completed with the slanted mirror stainless steel ceiling to give distorted mirror images and further enhances sculptural quality of the space, which projects strong visual impact through the full-height clear glass shop front.

EXIT出口

djs 綴飾

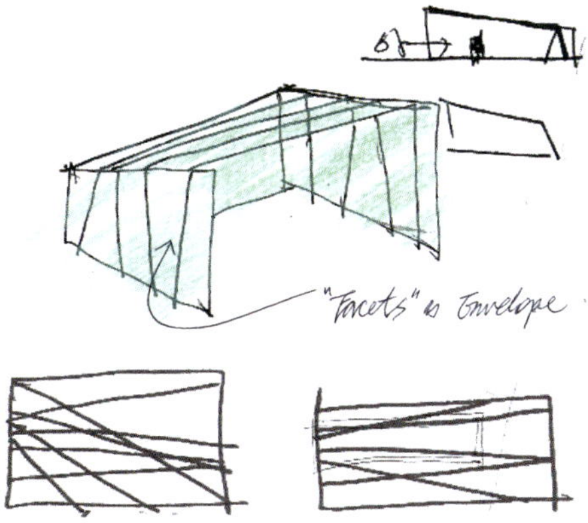
"Facets" as Envelope

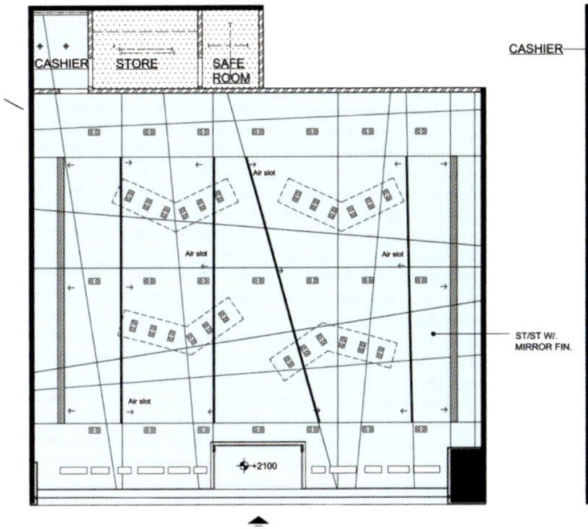
CASHIER
STORE
SAFE ROOM
Air slot
Air slot
Air slot
Air slot
ST/ST W/ MIRROR FIN.
+2100

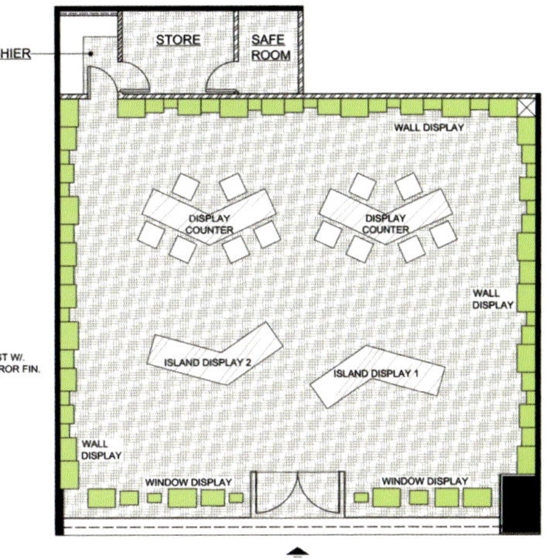
CASHIER
STORE
SAFE ROOM
WALL DISPLAY
DISPLAY COUNTER
DISPLAY COUNTER
WALL DISPLAY
ISLAND DISPLAY 2
ISLAND DISPLAY 1
WALL DISPLAY
WINDOW DISPLAY
WINDOW DISPLAY

琢面之美

本案在室内设计上，运用了钻石和翡翠不同形状的琢面和共同的性质、化学构造，来突出品牌的独特形象。为了达到这种效果，在室内设计上运用比喻手法，把购物空间演绎成一块宝玉。亮白色的条纹大理石壁橱搭配上白色大理石地板，为空间营造细微的基调；对称的布局与店铺的方正相呼应，再融合宝石琢面的几何造型，美观和实用兼备。亮白石壁橱有不规则的纵面，露出翠绿面。这种微妙的绿色点缀，道出了翡翠色泽玲珑剔透的自然之美。一排排浮动的玻璃陈列方块，大小、高度各有不同，别具格调，对顾客形成强烈的视觉吸引力。在方块内，有特制的磨砂亚加力展示架。店铺空间以棱角为整体风格，倾斜的镜面不锈钢天花板映照出扭曲的镜像，增添店铺空间的雕刻美，再透过店面落地玻璃，投射出强烈的视觉冲击效果。

Chinese Arts & Crafts

中艺品牌店

地点／中国香港　面积／234平方米　设计师／潘鸿彬
设计单位／泛纳国际设计顾问有限公司
主要材料／不锈钢、地砖、大理石

The interior design of Chinese Arts & Crafts (HK) Ltd. makes use of the metaphor of interpreting the shopping space as a "Chinese Gift Box". The dark spatial envelope of charcoal grey floor, and exposed ceiling are firstly set up to provide an understated backdrop to the space. This is followed by combining the rectilinear composition of grids in both two dimensional and three dimensional ways. In the Jewellery Zone, the feature wall-cum-ceiling of golden marble-patterned plastic fins create the texture of the "1st compartment" within the big gift box. Rows of golden yellow concealed light troughs reinforce the horizontality of this zone. Adjustable down spot lights create highlights to the delicate jewellery (diamond & jade) items with the help of the dim environment. In the Antique Zone, the grid display language becomes horizontal glazed shelves and glowing vertical dividers of the wall units. The central full height glazed "pavilion" display showcases becomes focus of the zone and provides opportunity for customers to look at the antique products in a three-dimensional way. In the Apparel Zone, the interplay of grid elements is finally completed with the glowing "wrapping paper" walls & ceiling. Custom made translucent monogram patterned glass diffuses golden ambient light to reinforce the nostalgic mood. System of free-standing and ceiling-suspended stainless steel hanger frameworks express the verticality and play the role of spatial dividers which penetrate into the apparel zone and float the merchandise to the eyes of the customers.

中艺品牌店的室内设计使用了"中华宝盒"的概念。围绕店铺的炭灰色地毯和天花为空间营造出淡淡的背景，然后从平面和立体上呈横直线融合方格结构。珠宝区的特色天花和墙身上有着金色大理石纹料突边，创造了大宝盒内的质感。金黄的灯槽增强了此区的横向感。在微暗的环境下，聚光灯突出了各式精致的珠宝。古玩区形成此区方格式的陈列，化为横式玻璃展柜，并成了闪亮的墙壁分隔线。中央的全高式玻璃展览亭是本区焦点所在，让顾客可从立体角度欣赏精致古玩。华服区方格元素的相互作用，最后通过发亮的花纸墙和天花完整呈现。玻璃上特制的半透明组合图案渗出金光，加强怀旧气氛。独立天花悬挂式不锈钢挂架系统，表现出垂直感，亦用做间隔空间，穿梭于精品区当中，把货品呈现在顾客的面前。

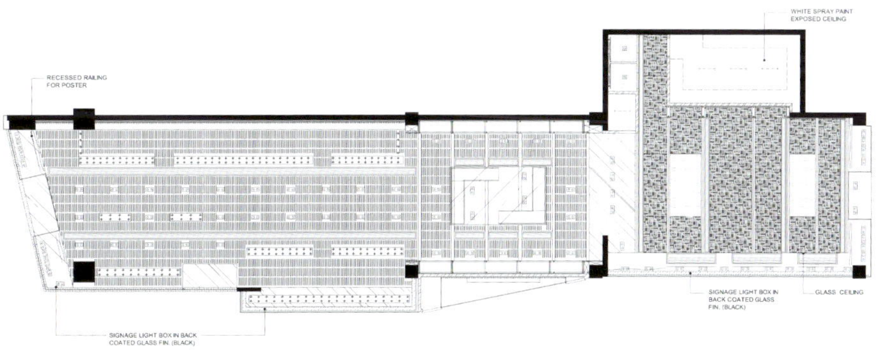

RECESSED RAILING FOR POSTER
WHITE SPRAY PAINT EXPOSED CEILING
SIGNAGE LIGHT BOX IN BACK COATED GLASS FIN. (BLACK)
GLASS CEILING
SIGNAGE LIGHT BOX IN BACK COATED GLASS FIN. (BLACK)

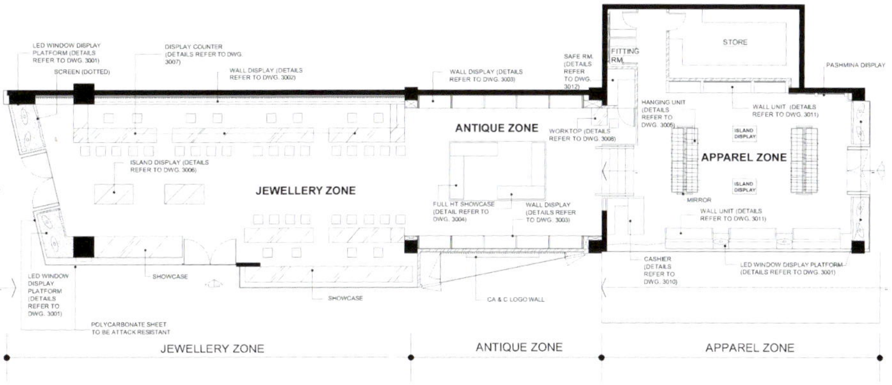

LED WINDOW DISPLAY PLATFORM (DETAILS REFER TO DWG. 3001)
SCREEN (DOTTED)
DISPLAY COUNTER (DETAILS REFER TO DWG. 3007)
WALL DISPLAY (DETAILS REFER TO DWG. 3002)
WALL DISPLAY (DETAILS REFER TO DWG. 3003)
SAFE RM. (DETAILS REFER TO DWG. 3012)
FITTING RM.
STORE
PASHMINA DISPLAY
HANGING UNIT (DETAILS REFER TO DWG. 3005)
WALL UNIT (DETAILS REFER TO DWG. 3011)
ISLAND DISPLAY
ANTIQUE ZONE
WORKTOP (DETAILS REFER TO DWG. 3008)
APPAREL ZONE
ISLAND DISPLAY (DETAILS REFER TO DWG. 3006)
JEWELLERY ZONE
ISLAND DISPLAY
MIRROR
FULL HT SHOWCASE (DETAIL REFER TO DWG. 3004)
WALL DISPLAY (DETAILS REFER TO DWG. 3003)
WALL UNIT (DETAILS REFER TO DWG. 3011)
CASHIER (DETAILS REFER TO DWG. 3010)
LED WINDOW DISPLAY PLATFORM (DETAILS REFER TO DWG. 3001)
LED WINDOW DISPLAY PLATFORM (DETAILS REFER TO DWG. 3001)
SHOWCASE
SHOWCASE
CA & C LOGO WALL
POLYCARBONATE SHEET TO BE ATTACK RESISTANT
JEWELLERY ZONE
ANTIQUE ZONE
APPAREL ZONE

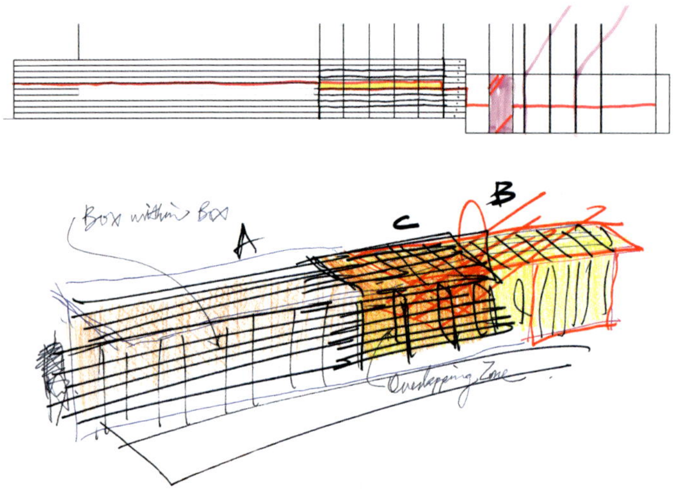

Box within Box
A
B
C
Overlapping Zone

YXZD Clothing Shop

"一席之地"服饰店

Temperament and Aesthetics of Space

This project is located in the commercial pedestrian mall on Zhongshan Road, and the brand is oriented to the urban consumer group of middle and young ladies, so the design style is leisure style. In the shop, black, white, gray, retro floor tiles, plain cement fiberboard and some classical and unvarnished materials are introduced to set off the characteristics of clothing style, and the balance and break of the island-style plan layout and the narrow space layout with a tubular partition make our guests first enjoy the visual feeling of looking without penetrability, separating without break and covering without block at the entrance. The purchase route among circuities increases the shopping fun, and the simple and unadorned space environment with slightly cool feeling shows out the simple and smooth lines, highlights that the environment owns the same temperamental aesthetics as the clothes, and emphasizes that the clothes are the protagonists in this space.

地点/佳木斯　面积/230平方米　设计师/王严民
设计单位/黑龙江省佳木斯市豪思环境艺术顾问设计公司
主要材料/复古地砖、素水泥密度板、镜品、白钢、复合地板、油漆

YXZD
YXZD

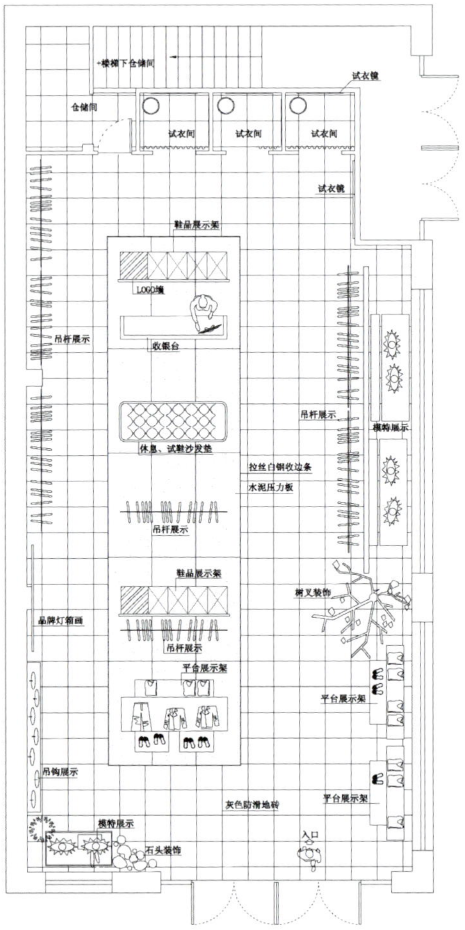
楼梯下仓储间
试衣镜
仓储间
试衣间
试衣间
试衣间
试衣镜
鞋品展示架
LOGO墙
收银台
吊杆展示
吊杆展示
模特展示
休息、试鞋沙发垫
拉丝白钢收边条
水泥压力板
吊杆展示
鞋品展示架
树叉装饰
品牌灯箱画
吊杆展示
平台展示架
平台展示架
吊钩展示
灰色防滑地砖
平台展示架
模特展示
入口
石头装饰

YXZD一席之地

YXZD

空间的气质美感

本案地处繁华的中山路商业步行街，品牌定位于都市的中青年女士消费群体，本案的设计风格为休闲风格。室内以黑色、白色、灰色与复古地砖、素水泥密度板和一些经典质朴的材料烘托服饰风格特征，并以岛式的平面布局平衡、分割较为狭长的筒状空间格局，让客人进入店内首先有种望而不穿、隔而不断、挡而不实的视觉感受，在迂回间的选购路线上增加了购物乐趣，在平实、略带酷感的空间环境内表现出简洁流畅的线条，强调环境与服饰一样具备气质美感，并强调服饰才是这个空间里的主角。

Stage for Hair
发舞台

地点/佳木斯　面积/320平方米　设计师/王严民
设计单位/黑龙江省佳木斯市豪思环境艺术顾问设计公司
主要材料/擦黑色曲柳板、水泥、马赛克、布艺、工艺玻璃、涂料

Let's go into the Stage for Hair with "primitivism" from modern city by the space-time tunnel. Stage for Hair subverts the public examining standard to hairdressing environment through the beauty of "new noble". The passion and magnificence perform at the same stage, the rough and textured cement is cool, the exquisite mosaic is gentle and beautiful, the silky and slippery fabric shows gentle beauty, and the European-style mirror frame, the crystal droplight demonstrate dignity in gorgeousness. The classical design vocabulary combining with fashionable designing elements is integrated into the elegant environment, so that our guests can experience a fire-new feeling on vision in the exotic culture and classical complex.

TONI&GUY
ACADEMY

穿越时空隧道从现代都市来到具有“尚古”之美的发舞台。发舞台以“新贵”之美颠覆了大众对美发环境的审视标准。激情与华丽同台献艺，粗犷质感的水泥酷感十足，细腻的马赛克温柔美丽，丝滑的布艺流露出阴柔的美感，欧式镜框，水晶吊灯，华美中彰显尊贵。古典的设计语汇结合时尚的设计元素融入到雅致的环境中，使客人在异域文化和古典情结中体验全新的视觉感受。

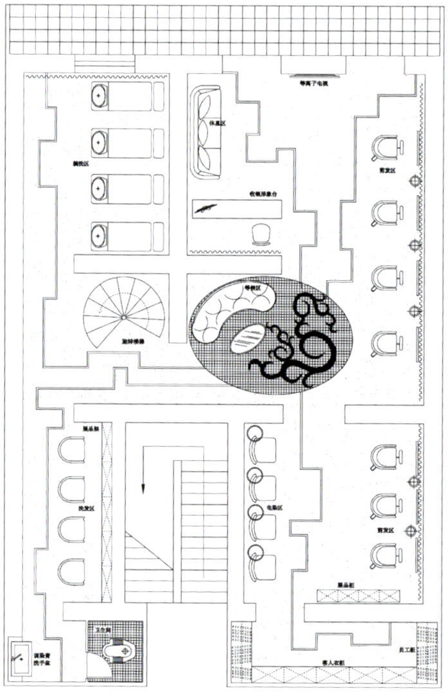

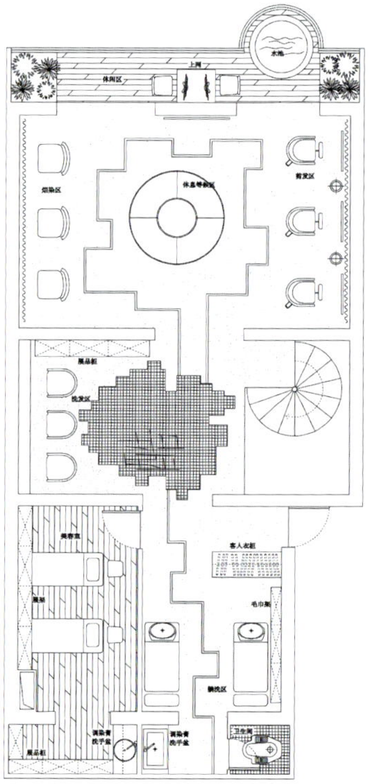

酷
HEILONGJIANG
JIAMUSI
SHISHANGDEFAWUTAI

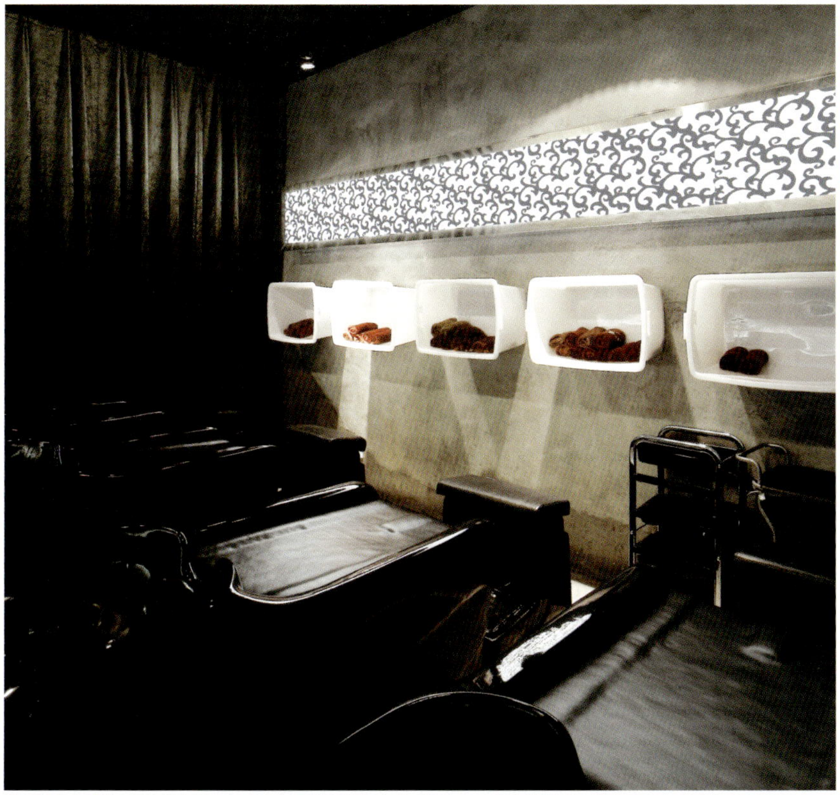

ROMANTICISM 2 in Hangzhou

杭州浪漫一身2

地点/杭州　面积/1140平方米　设计师/迫庆一郎
设计单位/北京卅口建筑设计咨询有限公司
主要材料/轻钢龙骨+密度板、不锈钢复合板、雪白色大理石、复合板、白色马赛克、白色水性涂料

Net Body / Skin

Clothes are that are cut and sewn with quadratic-element materials and then put on cubic-element body with three-dimensional packages and decorations, and space holds the post of packaging body and regulating body temperature. In the design concept of Romanticism 2, the partitions and accessories scattering among the space and clothes are considered to bring the effect of the "third layer of skin" into play, and "net"-shaped organic forms run smoothly throughout the space. From the sight of the façade, it looks like a "net" involved into the space, and then changes its forms naturally in the process of extension to wrap the whole first floor of the shop up. The "net" which is concentrated to be one bunch is outspreaded again after entering the underground space, and the same approach as the first floor is introduced here to extend the "net" to every corner. "Net" is changed fantastically to be partitions, the counter, chairs, accessories, handrails and other kinds of forms, and the "net" which plays the role of "third layer of skin" is made of "bones, meat and skin". "Bones" are the concrete reinforcing bars, "meat" is the plastic foam and glass fiber, and "skin" is epoxy resin and oil-based paint. The white wall of three-dimensional curved surface is drilled with a lot of holes for the display of samples, which expands the theme of body and clothes. In addition, the ceiling of the first floor is designed with mirror stainless steel on the base of achieving the effects of the visual impact, and it is designed to ensure the lighting of the shop and to overcome the shortcomings of the low ceiling.

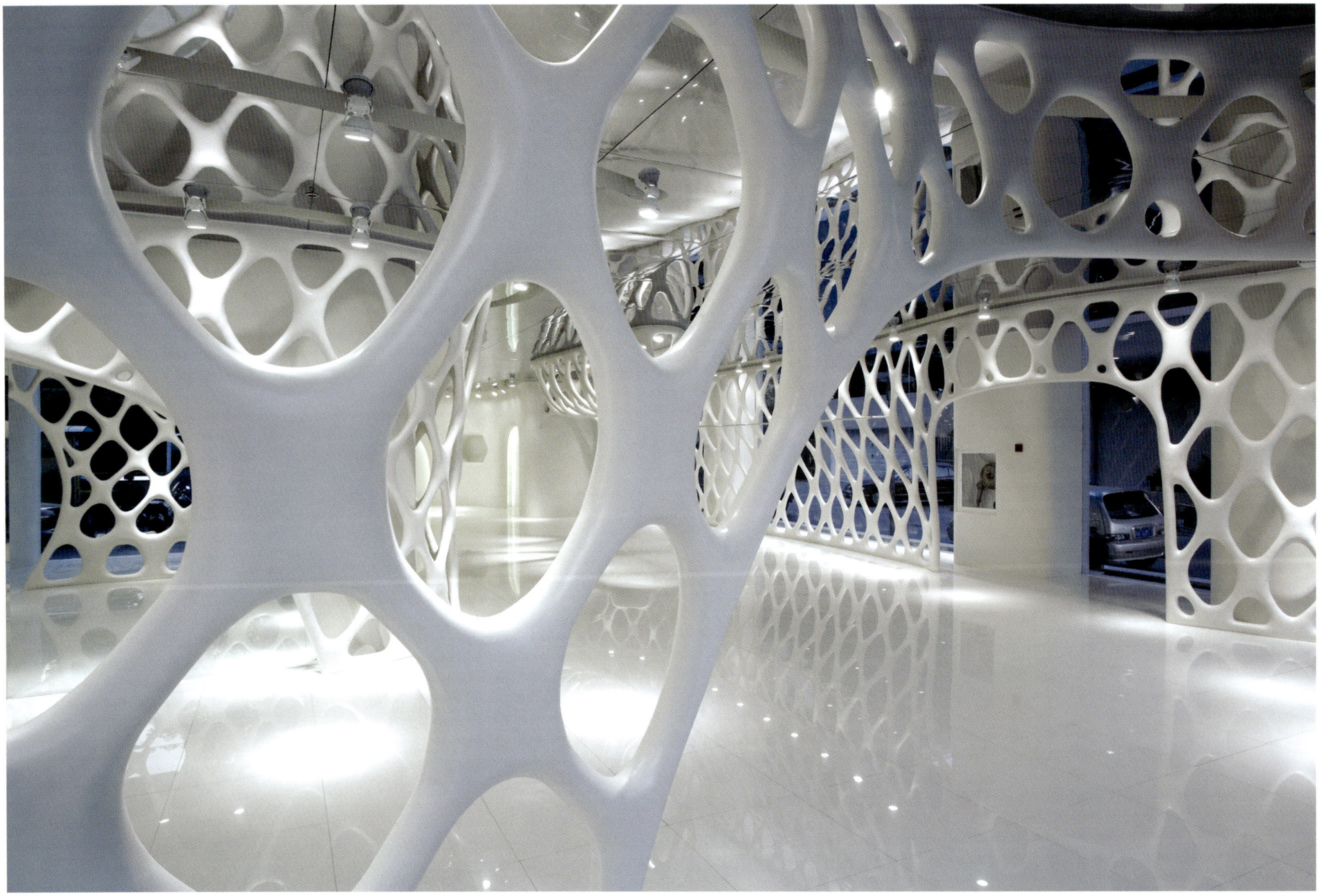

网 身体/皮肤

衣服是用二次元的布料进行裁剪和缝制，再将三次元的身体进行立体的包裹和修饰的东西。空间也担任着包裹身体、调节体温的作用。在浪漫一身2的设计理念中，设计师就考虑使分散在空间与衣服之间的隔断及器物发挥出“第三层皮肤”的功效。“网”状的有机形态流畅地贯穿于整个空间。从外立面看来像是被卷入到室内的“网”，在延伸的过程中自然地变化其形态，将整个一层的卖场包裹起来。在楼梯部分集结而成的一束“网”进入地下空间后再次被展开，与一层的做法相同，延续到各个角落。“网”幻化为隔断、柜台、椅子、器物及栏杆等各种不同的形态。起到“第三层皮肤”作用的“网”是由“骨、肉、皮”构成的。“骨”就是钢筋，“肉”是泡沫塑料及玻璃纤维，“皮”则是环氧树脂及油性涂料。在三维曲面的白色墙壁上，穿了很多用于展示样品的圆孔，扩大了身体和衣服的主题。另外，一层天花使用镜面不锈钢。这是基于达到视觉冲击的效果，保障卖场照度及克服天花较低的缺点而设计的。

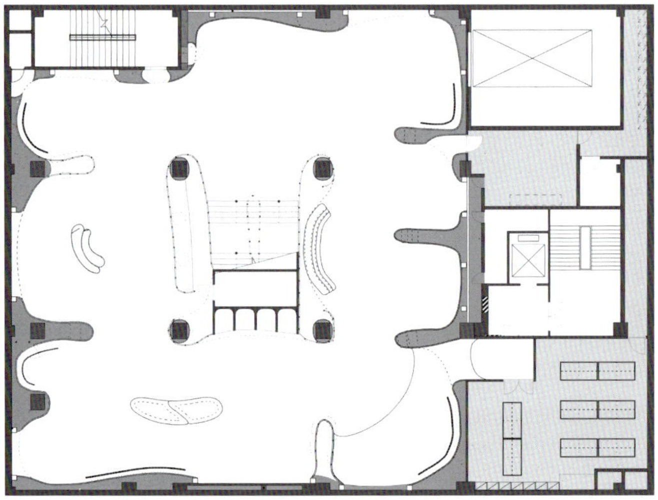

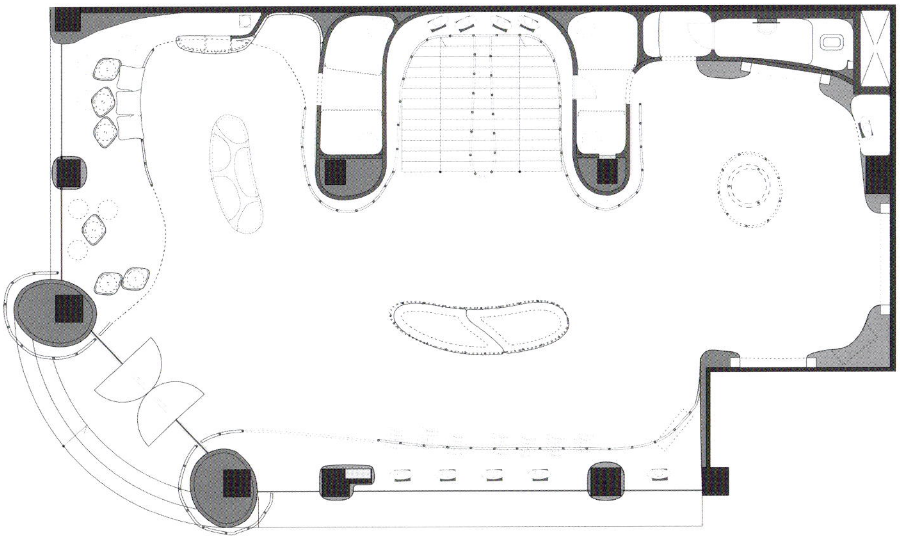

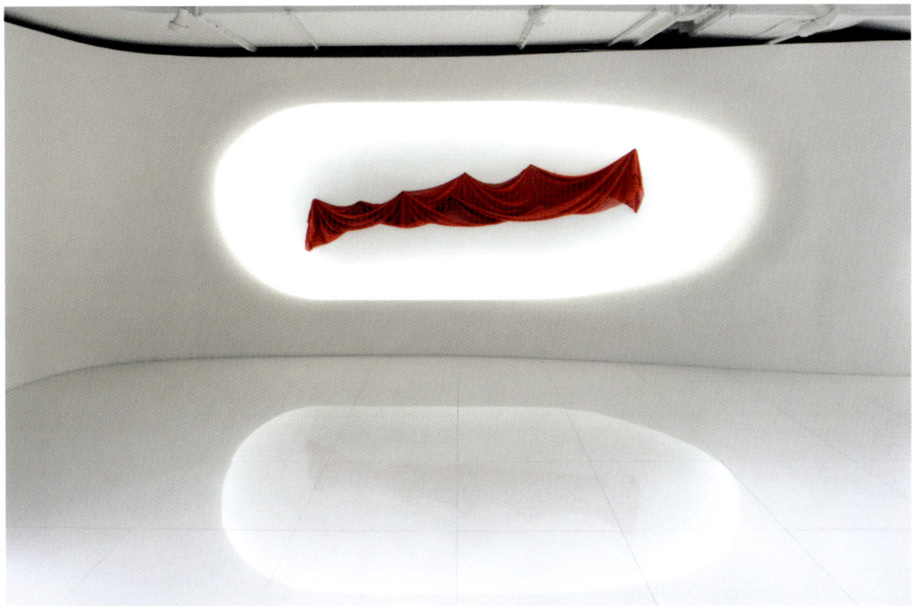

EIFINI in Chengdu

成都伊芙丽

地点/成都　面积/160平方米　设计师/迫庆一郎
设计单位/北京卅口建筑设计咨询有限公司
主要材料/白色瓷砖:600角（砂浆下地）、石膏板、不锈钢管

Circle Trajectory/ Levitation

In order to highlight the clothes, the ground, wall and ceiling are integrated with white color to remove the sense of existence of other elements in the shop. The cash desk and other furniture in the shop are designed with effect of simplicity as if they are the uplift of the ground tiles. A consistent "tube-style clothes hanging" which is totally 250 meters is introduced into such a space. The "tube" is supported by acrylic robs up and down on right and left, as if it is emancipated from the gravity, levitating with tridimensional twists in the space, so the space becomes broad without blocking the sight at the same time, and forms various scenes: the tunnel-like entrance formed with overlapping layers, spinning-down clothes show area metamorphosed into a whirlpool shape, a fitting room made of a hanging curtain, and square-like leisure space surrounded with tridimensional and multiple curves, which endow the space with richness and diversity. "Clothes" and "tube" play as the protagonist in turn, performing continuously throughout the shop. Every lady is sincerely invited into our EIFINI to stroll, as well as enjoy the beautiful melody changing by the seasons, as it were played by orchestra.

旋　轨迹 / 浮游

为了突出服装而将地面、墙面、天花用白色统一起来，以去掉店内其他元素的存在感。收款台等店内家具做出好像是地面瓷砖直接抬升起来的简约效果。在这样的空间中，让一根连贯的总长250米的"管式衣挂"介入其中。"管"由透明亚克力棒上下左右支撑起来，就像是从重力中解放出来一般，立体地缠绕着，浮游于空间内，使空间变得宽阔起来的同时毫不遮挡视线，并形成各个场景：重叠几层形成的好像隧道一样的入口、幻化成旋涡状旋转落下的服装展示区、挂起帘子变成的试衣间以及由立体的多重曲线环绕构成的广场式的休息空间，使空间获得了丰富性与多样性。"服装"与"管"交替着担任主角，在店内各处持续演奏。真心邀请每位走进伊芙丽的女性，徜徉于其间的同时感受这随季节而变化的宛如管弦乐团演奏般的优美旋律。

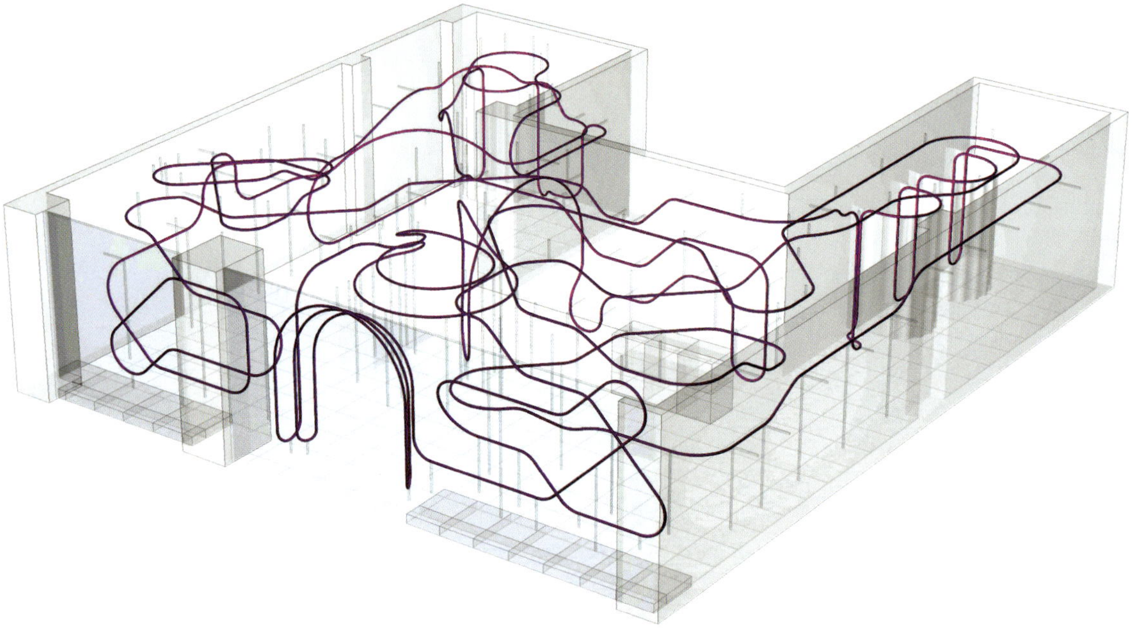

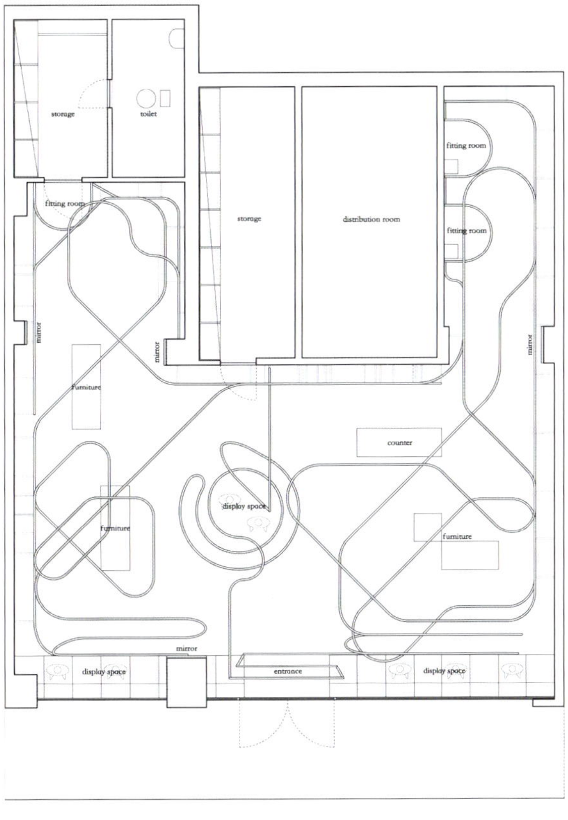
storage
toilet
fitting room
storage
distribution room
fitting room
fitting room
mirror
mirror
mirror
furniture
counter
display space
furniture
furniture
mirror
display space
entrance
display space

KID'S REPUBLIC in Shanghai

上海蒲蒲兰绘本馆

地点/上海　面积/105平方米　设计师/迫庆一郎
设计单位/北京井口建筑设计咨询有限公司
主要材料/人造大理石、白色涂料、刺杉原木、钢管、金属网、水泥砂浆

Tree Materiality / Fantasy

This project impresses us with a tree decomposing various elements—the tree stem, branches, cortices and leaves which are separately embedded in space. The processed elements remaining the sense of material and gaining the abstraction are used on the ceiling, walls, stairs and furniture, and then rebuilt in the space. The book shelves covering the whole wall come into view first, and the bookcases made of various timbers with different width, depth and thickness are three-dimensional forms, concave or convex, where there are cylinder bookcases made of hollow-out stems to bring new "face" into the space. The floor reflects the luxuriant leaves, as if we can really feel the sunshine through the branches and leaves. The ceiling is designed with logs with branches equidistantly, and from the street, we can see the growth ring on one end of trees through the glass. Once entering the second floor, we can see the hemispherical activity space covered with rough cortices. Inside is a pure white space like a snow room, where the ceiling is designed with irregular circular holes. The floor is designed with different levels as a stage and an auditorium, and children can read while lying on the colorful carpet.

没有画的画册
ABCDE
FGHIJK

在森林里
洗澡
咿-哦！
别叫了！
玛德琳卡的狗
Peter Sís
要是你给老鼠吃饼干

森林里
的聚会

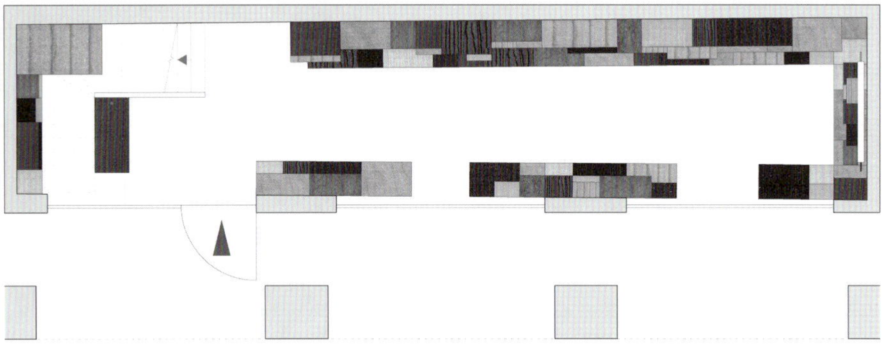

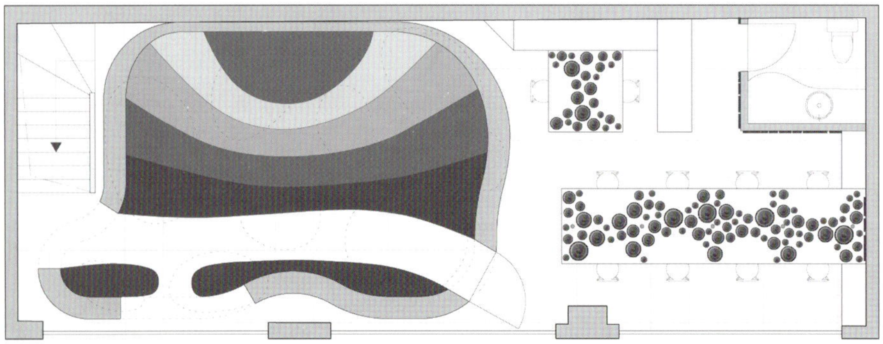

树　物质性 / 幻想性

本案给人的印象是一棵大树分解出各个要素——树干、树枝、树皮和树叶，分别嵌于空间中。将加工后既保有物质感又被抽象化了的这些素材用于天花、墙壁、楼梯和家具，由此在空间中被重新构建。首先映入眼帘的是立于一整面墙壁的书架，使用各种树木的集成材做出宽度、深度、厚度各异的书柜，它们或挑出去或收进来，是一个立体构成。其中又有利用一段树干将其挖空变成的圆筒书柜，以此带来了新的"表情"。地板上投影出繁茂的树叶，好像真的能感受到透过枝叶泄露出的阳光一般。天花上等间距地排列着带有枝杈的圆木，从街道上走过，透过玻璃可以看见树木一端的年轮。一进二楼，便可看到由粗砺的树皮覆盖的半球形的活动空间。内部是纯白的好像雪屋一样的空间，顶上开有几个不规则圆形的孔洞。地面做出高差，作为舞台及观众席来使用，孩子们可以躺在彩色的地毯上阅读。

布鲁姆博士
搞不懂
挖鼻孔
小泰的小猫
三只山羊
快活的狮子
好奇的乔治

雪山救助犬和旅行者

紙袋公主
小魚散步
快樂的一天
天才大笨貓
雲上的小孩

BRANCH in Changchun
长春藤

Nervation Divarication/Integration

This project is a facility for cultural exchange aiming at introducing Japanese culture, which is arranged and established by International Communication Fund. Trees are chosen as the theme of the design, and all kinds of facilities surrounding the space own the same allure as a forest to attract us to the space. We can sit or lean on the rattan branches, or find a hidden place through the branches. It is a space that can evoke our imagination, and each part is designed with a davenport, a sofa and a table which are designed according to our body size to facilitate the use. The floor, wall and ceiling are mainly white, and the original columns are packaged with stainless steel mirrors to play down the columns. The green-based facilities bring an effect of leisure into the whole space, and the gradual changing of cool colors and the LED lights hidden under the "branches" highlight the organism of nervation. The spherical LED lights placed everywhere means new green enjoying the dew and showing out. The vivid divarication and integration of the tree are the symbols of Sino-Japanese exchange and growth.

地点／长春　面积／280平方米　设计师／迫庆一郎
设计单位／北京卅口建筑设计咨询有限公司
主要材料／石膏板、白色瓷砖:600角（砂浆下地）、不锈钢

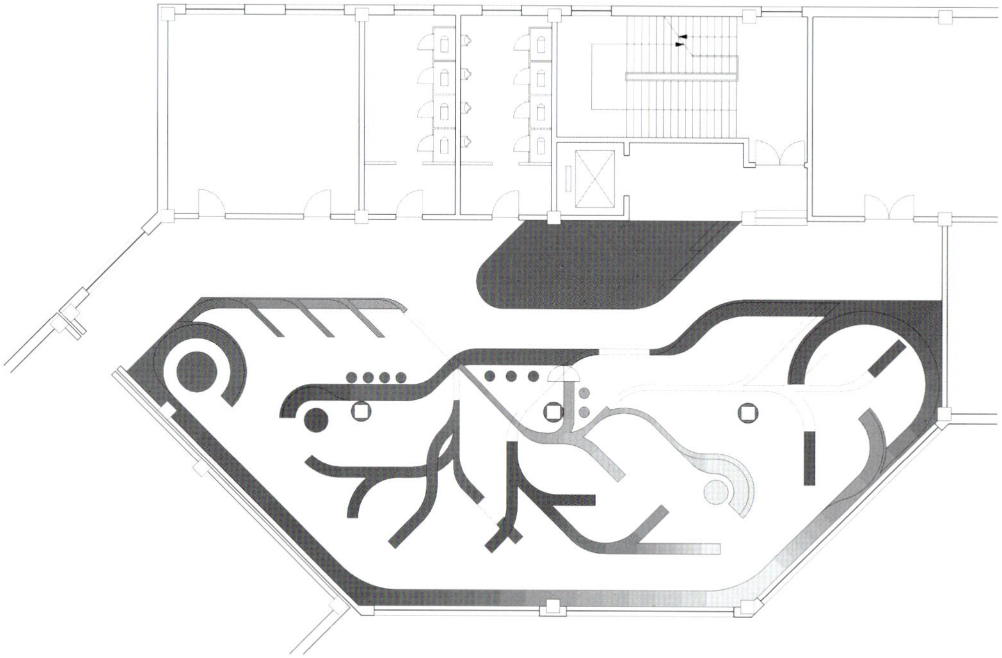

脉络　分歧 / 合成

本案是一个旨在介绍日本文化的由国际交流基金筹备设立的文化交流设施。设计师以树木为主题进行了设计。环绕于室内的各种设施的脉络有着像森林一样的诱惑力，吸引人们进入。人们或坐在藤枝上或倚在藤干上，又或者在枝杈的间隙间发现一个隐蔽的场所。这是一个能够唤起人们想像力的空间，各部位都是依据人体尺寸设计的使用起来非常方便的长椅、沙发、桌子。地面、墙面、天花以白色为基调，原有的柱子用镜面不锈钢包裹起来，以此淡化柱子的存在。以绿色为主的设施给整个空间带来放松的效果，冷色系的渐变和埋在“树枝”下的LED灯带强调了叶脉的有机性。安放在各处的球形LED灯寓意沐浴着朝露而焕发出的新绿。这棵树生动的分歧与合成，象征着中日两国的交流与成长。

FLATFLAT in Harajuku

原宿FLATFLAT

面积/330平方米　设计师/迫庆一郎
设计单位/北京卅口建筑设计咨询有限公司
主要材料/砂、霓虹灯、涂料

Park Reality / Supposition

"FLATFLAT in Harajuku" is an experience shop of NHN Company in Japan. In the shop, our customers can experience the services on "Hangame", the net of a online game community, and "hange.jp", the net of a mobile phone game, which are operated by this company. This show flat is a lathy space which is 3.5m wide and 45m long, at the same time due to repeated additions it forms complex plane and structure. "FLATFLAT in Harajuku" puts forward a concept of "Future Park", that is the real space where anyone can easily enter and live free is collocated with suppositional elements. It is a space where "organic wet" and "bleached dry" coexist, which means the modern society operating on real and suppositional meanings. The "organic wet" composed of curves owns a cave-like space feature, and with the scene unfolding, our customers can find the most suitable place in accordance with personal purpose or feeling. The curves which are designed on the base of body size form an organic style, which leads us to go automatically closer and closer. At the same time, the "bleached dry" which is composed of the inorganic textures of the white wall, neon lights on the ceiling and cement mortar ground produces a suppositional feature. In order to prevent the cement mortar ground from cracking, the ground is deliberately designed with apertures, and the lines of neon lights are consistent with the curves of the wall.

ふらっときてね。
bb flat flat
HANGame square

公园　现实／虚拟

“原宿FLATFLAT”是NHN Japan公司的一家体验店。顾客可以在店内亲身体验这家公司所运营的在线游戏社区网站“Hangame”及手机游戏网站“hange.jp”上的服务。本案是一个幅宽3.5米、进深45米的极为细长的空间，同时因反复的增建形成了复杂的平面及结构。“原宿FLATFLAT”提出了“未来公园”的理念，即在任何人都可以轻松进入且自由活动的现实空间里配合上虚拟的要素。这是一个“有机的wet”与“被漂白过的dry”共存的空间，表示着在现实与虚拟两义性上运营的现代社会。由曲线构成的“有机的wet”具有洞窟般的空间特性，随着场景的不断展现，顾客可根据个人的目的或心情找到最适合自己的地方。基于人体尺寸而定的曲线形成有机的形态，引导人们自然而然地接近和亲近。同时由白墙、天花的霓虹灯管、水泥砂浆地面的无机的质感构成的“被漂白过的dry”产生出虚拟的特性。为了防止水泥砂浆地面出现裂缝而故意做出的缝隙，以及霓虹灯的线条都与墙面的曲线一致。

Goods
Free Game

Cafe
Touch Panel
Free Game
PLAY IT NOW

Cafe
Touch Panel

ふらっときてね。
bb flat flat
Communication

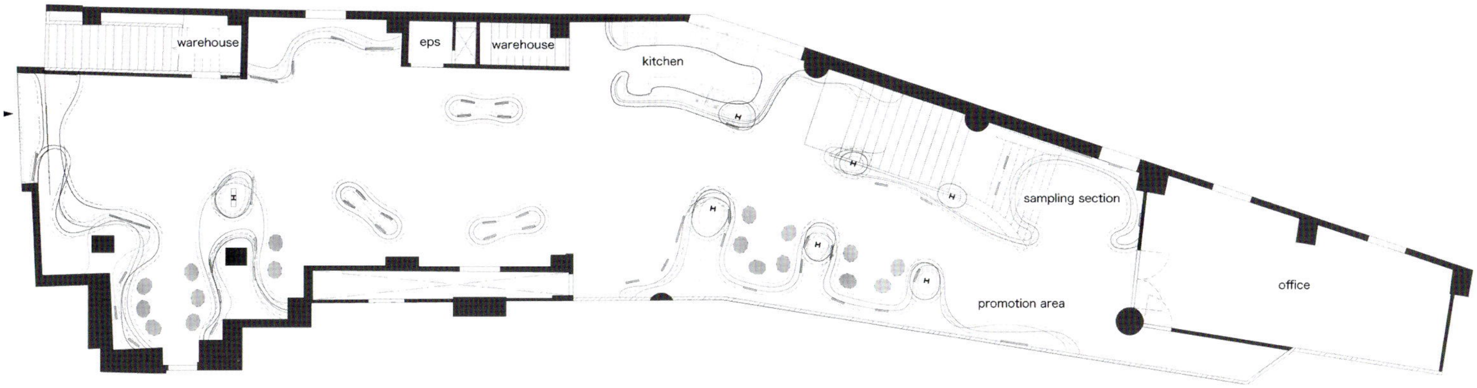
warehouse
eps
warehouse
kitchen
sampling section
office
promotion area

IMAGINARIUM in Barcelona

巴塞罗那IMAGINARIUM

面积/990平方米　设计师/迫庆一郎
设计单位/北京卅口建筑设计咨询有限公司
主要材料/石膏板、蓝色水性涂料（亚光）、白色水性涂料（亚光）、木板、创作壁纸、白色大理石（光泽）

Cave Contour Lines/ Isochromatic Lines

This project is a long and thin space which is 6 meters high and 28 meters deep with three floors. The cave is the element that just meets the requirement: we can not have a full picture of the panoramic view at one sight, however, with the penetration of the space, the full picture show in front slowly and continuously. The rainbow is introduced into the space to highlight the commodities and create an enthusiastic atmosphere. The colors of the rainbow change gradually with the vertical increase of height. Each of the six levels of the show cabinet highlights smooth curves, which seem like the contour lines to reproduce the natural terrain. Some tongue-shaped extrusive show cabinets can meet the needs of all kinds of functions, such as the exhibition platform, partition signs, tables, chairs, the counter, book shelves and partitions. Various kinds of commodities are scattered in a three-dimensional space as a result of the different shapes of each display rack. When we look down from the top, we can see the display racks with various kinds of colors like isochromatic lines laying out the distribution of the commodities, at the same time highlighting "stress of space" .

Servicio
Especial
Coleccionistas

Cool boy

WOLVES
THE SNOWY DAY
Opposite

ZOO
GONE WILD
365
ItsImagical

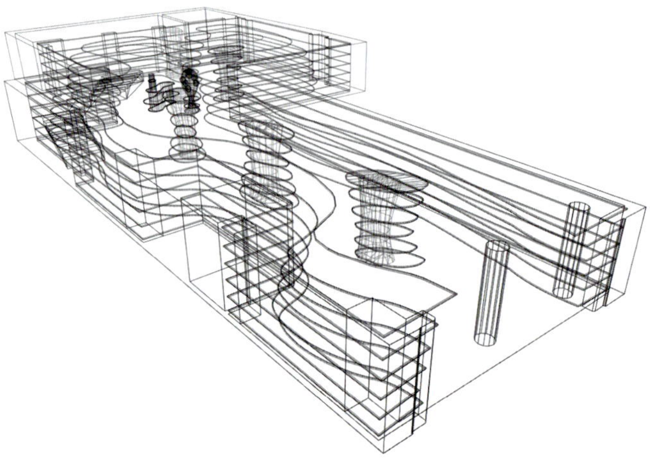

洞　等高线 / 等色线

本案是一个6米宽、28米进深的细长空间，共有三层。洞窟是正好符合要求的元素：一眼望去并不能将全部景象尽收眼底，而是随着空间的深入，整体景象慢慢连续地在眼前展开。设计师在空间中使用彩虹来突出商品，并营造热烈的气氛。彩虹的颜色也在垂直方向随高度的增加而逐渐发生变化。6层陈列柜的每层都强调了圆滑的曲线，好像等高线一样再现了自然的地形。一些像舌头一样突出的陈列柜可以满足促销产品的展台、区域标志、桌子、椅子、柜台、书架或者隔板等各种功能性需求。由于展架每层形状不同，各种商品被分散地分布在三维空间中。由上而下俯视时，可以看到各种颜色的展架好像等色线，勾画出了商品的分布，同时也强调了“空间的应力”。

Camelot

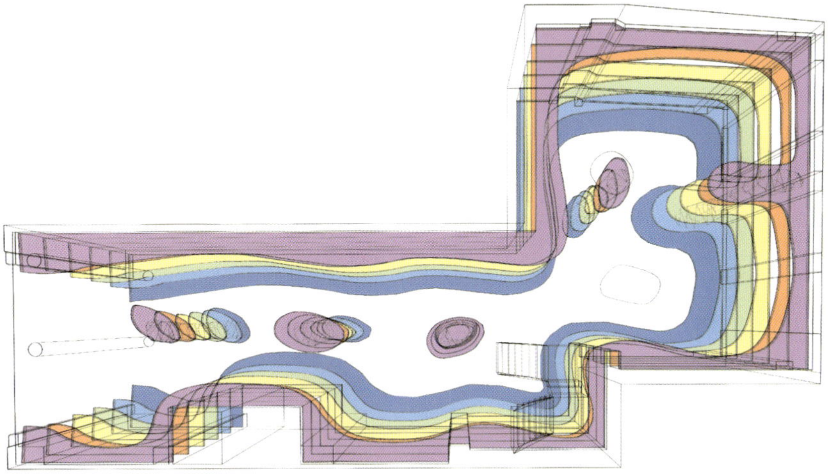

GOBI in Ulaanbaatar

乌兰巴托GOBI

面积/1150平方米　设计师/迫庆一郎
设计单位/北京卅口建筑设计咨询有限公司
主要材料/不锈钢、地砖、大理石

Regeneration Smoothing / Rough

It is a design of the factory direct store of GOBI Company, which is the largest senior sweater brand in Mongolia. Our designer finds a deserted old warehouse in the building site, and then is interested in using large space actively for the coexistence of substantial inventory of stores and the open shopping mall, so the warehouse is renovated. The arrangement of concave-convex bricks highlights the unique sense of roughness in the old warehouse. Other broken parts are not repaired, and both inside and outside are painted white, like the accumulation left by time. In order to form a contrast with the original rough constructive body, the interior partitions are decorated with smooth glass. The shopping mall is encircled with glass and wall furniture, and a huge number of clothes are stored on the fringe of the mall in order to bring the customers with sense of sureness through such kind of cluster round. The curve wall which is full of changes is made to attract customers as if they are walking on the street, and it is linked with the glass show window cantilevered by overhanging beams. The glass show window hanging in the air which symbolizes the modernity and the rough and simple warehouse contrast and coexist with each other.

GOBI
FACTORY STORE

NEW STYLE
FACTORY

再生　平滑 / 粗粝

这是对蒙古最大的高级羊毛衫品牌GOBI公司工厂直营店的设计。设计师注意到建筑用地内一个荒废的旧仓库，于是对活用大空间以实现大量库存品与开放性卖场的共存产生了兴趣，便对仓库进行了改造。设计师通过码出凹凸的砖，强调了旧仓库特有的粗犷感。其他的伤处也并不修补，内外都只是涂成白色，做出好像时间的积淀残留下来的一般。与原始粗糙的建筑躯体形成对比的是，室内的隔断多使用平滑的玻璃。卖场被玻璃及墙面家具包围，在其外侧储存有数量庞大的衣服，旨在通过这种簇拥为顾客在购物的时候带去一份踏实感。做出富有变化的曲面墙壁，好像走在街巷里一样，吸引着顾客前行，并与靠挑梁悬挑而出的玻璃橱窗相连接。悬浮在空中的玻璃橱窗所象征的近代性，与有着粗粝光景的仓库对比地共存着。

GOBI
FACTORY STORE

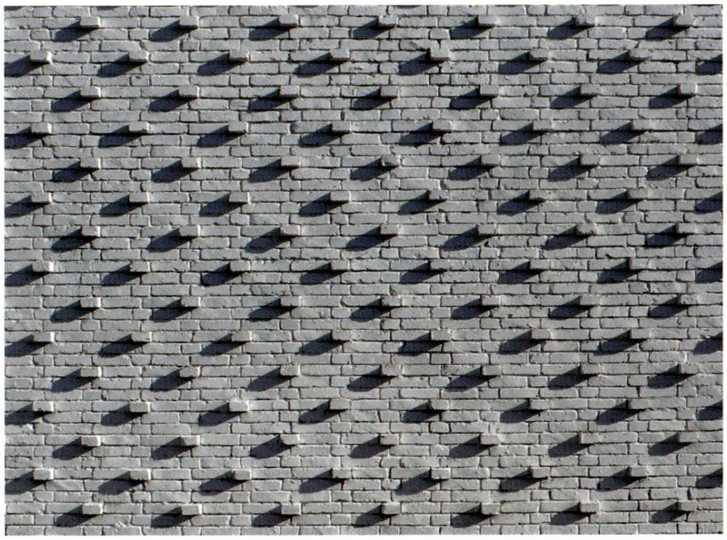

GOBI FACTORY STORE

PAMPER HEiRESS

PAMPER HEiRESS

地点/中国台北　面积/132平方米　设计师/陆希杰
参与设计/曹车政、刘致佑、郑家皓　设计单位/CJ STUDIO
主要材料/盘多魔、水泥漆、马来漆、镜面不锈钢、超白强化清玻璃、泰丝
摄影师/Marc Gerritsen

This show space is designed for PAMPER HEiRESS, a clothing brand. Clothing and jewelry are on display in this space, which aims to present a bright and sparkling ambiance. The theme "Diamond House" is adopted for the fashion brand Pamper Heiress' display place. In this diamond concept, the vision focus is a glass showcase, which is cut and segmented to imitate the multi-angle light reflections of diamonds, with various diamond shaped platforms on the four sides to conform to it. The ceiling is decorated with folded lines and shapes corresponding to the angled jewelry showcase for a totally consistent look. The images of diamonds are everywhere, such as the magnified "diamond-ring stand" hooks in the fitting rooms. The whole area design is neat and smooth from large items to fine details, creating a sophisticated space.

为服装品牌PAMPER HEiRESS打造的展示空间，空间内同时陈列服饰与珠宝，并期望呈现闪耀亮丽的空间氛围。以"Diamond House"作为发展主轴，以钻石造型为概念，做玻璃的切割拼接，呈现钻石的多角度反射而形成的珠宝展示柜为空间中的视觉焦点，四周中岛平台则以切割成不同造型钻石为概念，与之呼应。为了与角状的珠宝柜台搭配，天花设计以折叠的线条与曲面处理，从而达到整体空间的协调感。空间中处处皆有钻石的意象，更衣室的挂衣钩以放大版的钻石戒座为概念，整体空间简洁流畅，由空间至细节都精细处理，充分营造精致的氛围。

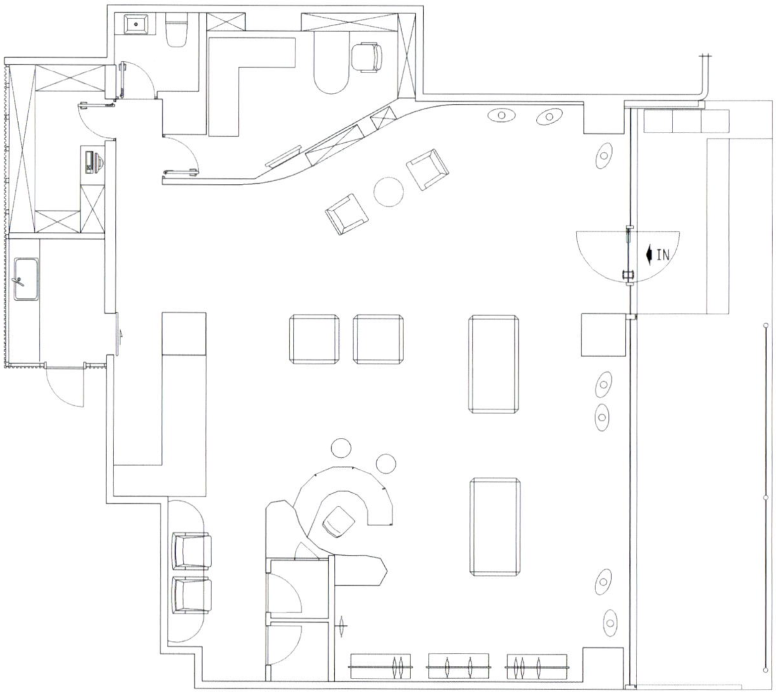

Giant Bicycle Inc.'s Flagship Store

捷安特女性旗舰店

地点/中国台北　面积/284平方米　设计师/陆希杰
参与设计/曹车政、刘致佑、张森贺、陈宥诠、郑家皓
设计单位/CJ STUDIO
主要材料/毛丝面不锈钢、雾面石英砖、马赛克、人造皮、卷帘
摄影师/Marc Gerritsen

This retail store is constructed as Giant Bicycle Inc.'s flagship store specifically for female customers. A "Landscape Jigsaw Puzzle" is the central concept of the floor design, with each piece representing a different recreation scene to present the change of one's mood during different activities and how they are linked together. Such concept is incorporated in the spatial design for the store and is portrayed through the combination of furniture and the ceiling design. The ceiling is decorated in streamline shapes that resemble clouds to suggest themes of nature. As well, it also serves as an extension of the landscape jigsaw puzzle pieces. The furniture consists of irregular shaped sofas made from chocolate-colored cowhide material to create randomly shaped landscapes through the utilization of various patchwork techniques. The focus of furniture design rests on the planning of a multifunctional space that could also function as a convenient yet lavishly decorated environment for recreation so that the store would be more than a dull place that sells bicycles. On the first floor, the retracted shop front looks just like the frame of a stage; the ambient lighting along with the indoor trail lights renders a well-lit visual space. The display area features bicycles and relevant merchandise exhibited over quadrilateral frames of varying sizes like a museum. Such display style not only makes browsing much easier for the patrons, but also offers a more refined presentation of the featured products.

这是一间针对女性使用者的捷安特女性旗舰店，"地景拼图"是设计的主要概念。拼图代表着各个休闲场景、活动的心情转换与连结，将此概念融入空间，表现于家具组合及天花中。天花以流线、自然有机的造型刻画出白云，亦是地景拼图的延伸。家具则是以巧克力色牛皮材质做不规则地景貌的沙发，融入各种拼贴组合手法。设计重点是规划多机能的使用空间，同时也做出精品质感的休闲便利环境，不只是卖场而已。一层部分，退缩的门面让入口处有如舞台的框景，将阳光引入室内，搭配室内轨道灯光，营造明亮的视觉空间效果。展示陈列部分以不同大小的方格摆放脚踏车及商品，如同博物馆中的陈列方式，这样的陈列易于观赏，也提高了商品的精致度。

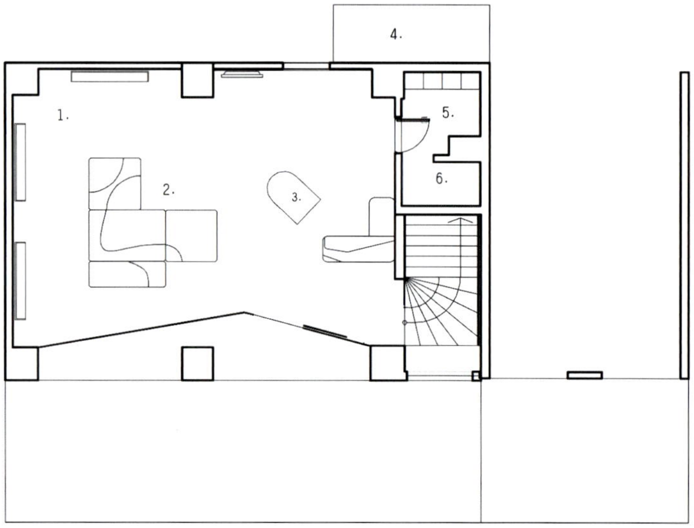

一层
1.展示区
2.展示区
3.电梯
4.洗车区
5.产品存放区
6.员工室

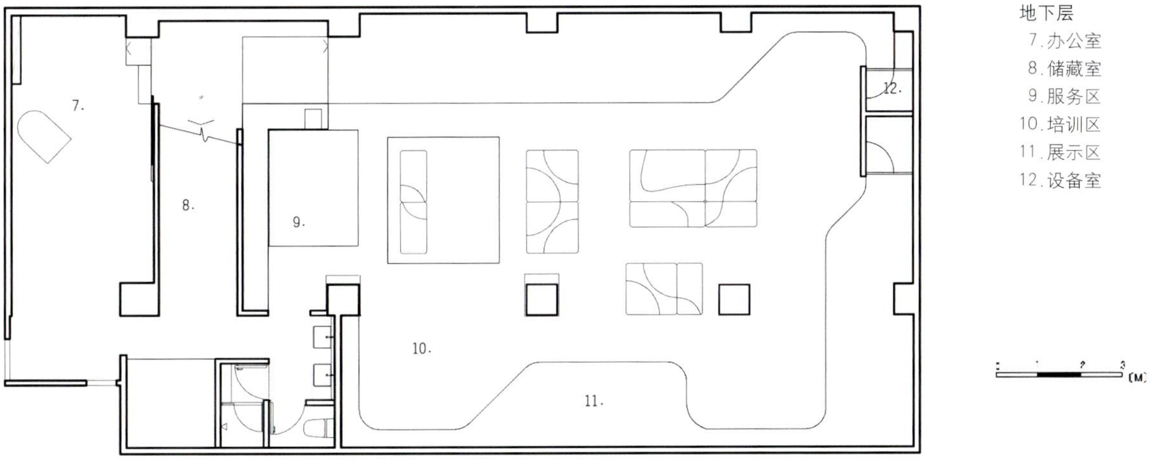

地下层
7. 办公室
8. 储藏室
9. 服务区
10. 培训区
11. 展示区
12. 设备室

維修價目
車輛調整諮商
車輛外觀美容
車輛保養
組車服務/台
車輛運費
配件組裝/件

FZL Concept Shop

FZL概念店

In this project, black and white are chosen as the main tone, the modern style is applied as the image, and the design of the concept shop continues the original colors of black and white with European-style, palace and exquisite ideas. The sequence and composition of black and white form the sequence of the design, and the space is endowed with strong sense of sequence; the design is interspersed with the organic decorations, and all of these impact us strongly on vision. The lively lighting, clear and penetrative glass, mirror stainless steel which makes the space scatter and change, elegant and fashionable wallpaper with black and white European-style patterns, a stocky soft package of black leather, Roman decorative columns, a series of elements form the design theme of the designers—chess in the modern palace.

地点/杭州　面积/150平方米　设计师/林森、谢国兴
设计单位/杭州肯思装饰设计事务所
主要材料/黑色强化地板、镜面不锈钢、欧式花纹壁纸、黑色皮软包、黑白色抛光砖、罗马装饰柱

E|Z|L

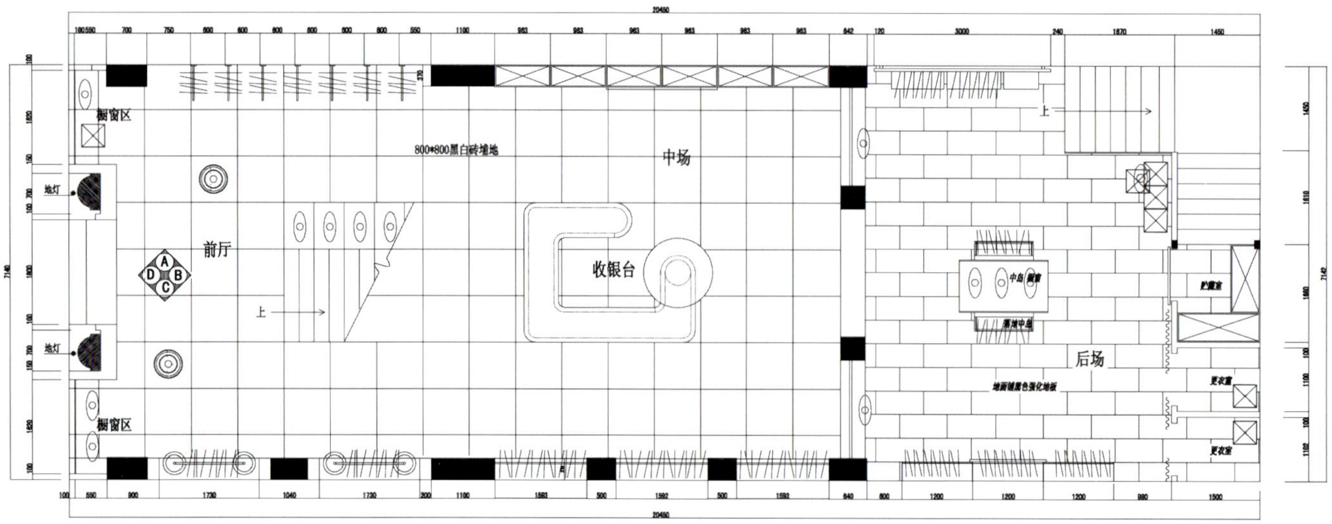

橱窗区
800*800黑白砖墙地
中场
地灯
前厅
收银台
上
后场
更衣室
橱窗区

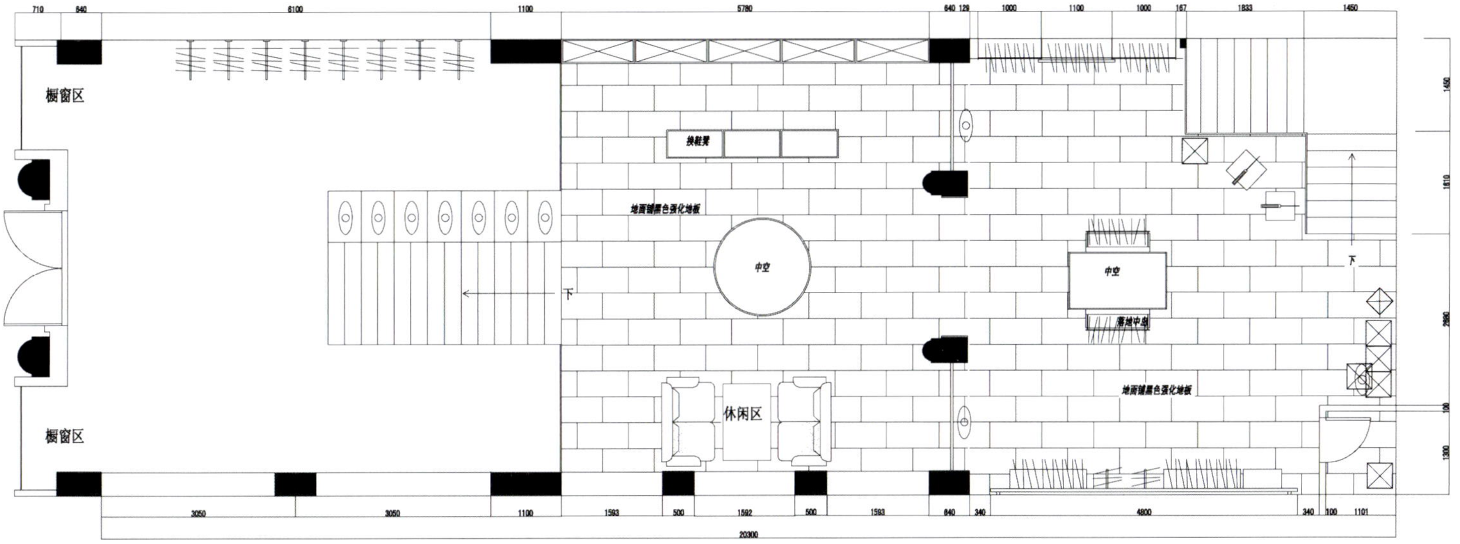

橱窗区
换鞋凳
地面铺黑色强化地板
中空
下
中空
落地中岛
休闲区
地面铺黑色强化地板
橱窗区

木案以黑白为基调，以现代风格为形象，在概念店的设计中延续了原有的黑白色调，并加入了欧式、宫廷、精致的创意。黑白色彩的序列、组成形式的序列，设计师将空间赋予极强的序列感；有机的体块穿插，这一切都给人极强的视觉冲击力。活泼的灯光、清透的玻璃、使空间错落变化的镜面不锈钢、优雅时尚的黑白欧式花纹壁纸、敦实的黑色皮软包、罗马装饰柱，这一系列的元素组成了设计师的设计主题——现代宫殿中的国际象棋。

Bristol Hair Salon

布里斯托美发沙龙

地点/深圳　面积/152平方米　设计师/Jacky
参与设计/Viky(汤伟奇)、Cat(吴小梅)、David(吴振飞)、Jenny(詹娟珲)
设计单位/香港皇朝室内设计有限公司、深圳市中港装饰设计工程有限公司
主要材料/明镜、马赛克、高级金镜、艺术涂料、大理石、透光石、芬兰木、文化砖、仿古砖、简一地板砖、仿条形实木地板砖

Bristol Hair Salon is the first commercial space with Southeast Asian style as the concept design in Hong Kong-Shenzhen area, and it is a SALON providing beauty services professionally for top-grade customers, which is located on the first floor of Building B in the World Financial Center. Southeast Asia is located in the southeast of Asia, which connects China in the north and Australia in the south, and the main countries include Thailand, Malaysia, Singapore, Philippines, and Vietnam and so on. The design of the space makes us feel the tropical climate in Southeast Asia, at the same time it shows us Buddhist culture of Thailand. The salon is designed with a unique VIP room, a classical reading bar, a leisure drink bar, an elegant dye and perm area, Thai-style shampoo area, reception area and so on, and each exquisite detail shows the personalized design.

布里斯托美发
SALON
布里斯托美发 SHARON

布里斯托美发沙龙是深港地区首家以东南亚风格为概念设计的商业空间，店址位于世界金融中心B座首层，是一家为高端客户提供美容美发专业服务的沙龙。东南亚位于亚洲的东南部，北接中国，南望澳大利亚，主要国家有泰国、马来西亚、新加坡、菲律宾、越南等。空间的设计让人感受到东南亚的热带气候，同时又能流露出泰国的佛教文化。沙龙设有独创的VIP室、经典书吧、休闲水吧、高雅烫染区、泰式洗头区、客户接待区等，每一细微之处均显示出人性化的设计。

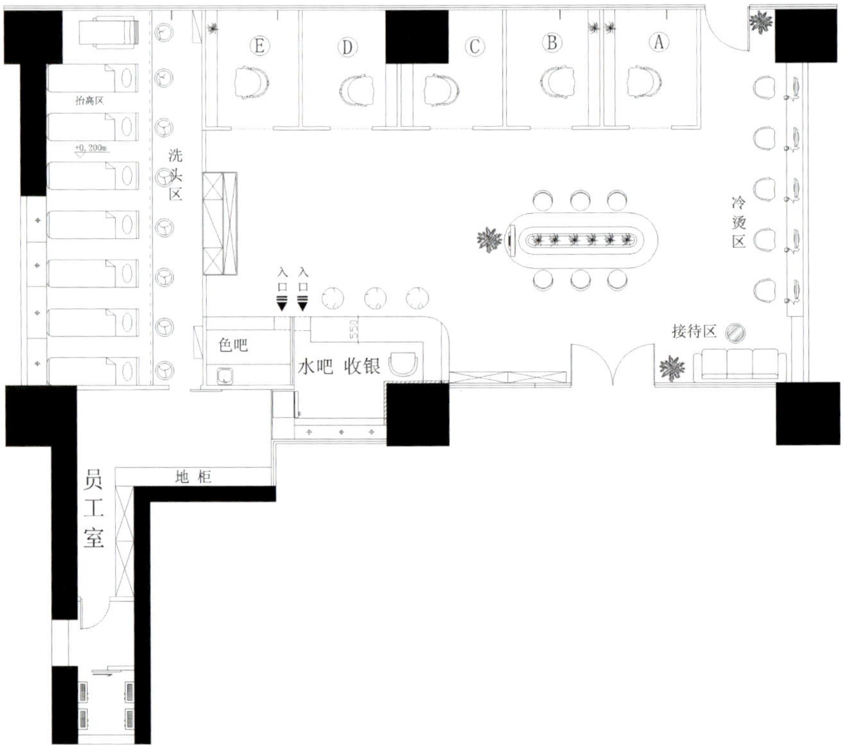

Yogo Juice in Jinmao, Shanghai

Yogo Juice上海金茂店

Yogo Juice, which is called Juice Buster in North America, enters into China with hope of a fire-new image, so it is designed by the designers from Sigdesign. The designers hope that the new juice shop can be exquisite and simple, natural and fresh with sense of science and technology, so they call this style "scientific and technological nature". All the walls are made of aluminum sheets in pursuit of the "ultrathin" science and technology concept, including ceiling design, the bar counter and the menu, and we can imagine tearing an ultrathin paper, and then folding it. The heightened bar printed with patterns of green leaves, small round tables, and the smooth green wall stickers make you feel that you can really enjoy the natural air.

地点/上海　面积/88平方米　设计师/李睿、曹秋月
设计单位/上海形格艺术设计有限公司
主要材料/铝板、烤漆、面喷图案

Yogo

Yogo
Order Here

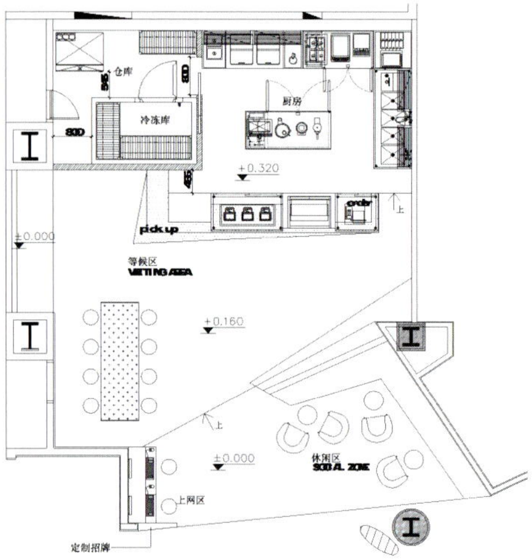
仓库
冷冻库
厨房
800
+0.320
pick up
order
上
±0.000
等候区
WAITING AREA
+0.160
上
±0.000
休闲区
SOCIAL ZONE
上网区
定制招牌

Yogo
绿色有机饮品
天然碱性，
提高免疫！
Fresh-squeezed
Sunlight!
Incredible green wonders of
Yogo wheatgrass!
把阳光"新鲜现榨"！
常青魔力素
给你源源不断的
纯净能量
和青春活力！

Yogo
YOGO Smoothie & Juice Bar

Yogo Juice，在北美叫做Juice Buster，进入中国后希望有一个全新的形象，于是找到Sigdesign的设计师。设计师希望新的果饮店能够精巧却又朴实，有科技感但也很自然清新，于是他们给这种风格取名为“科技自然”。所有的墙面都采用铝板制作，以追求“超薄”的科技概念，包括顶面造型、吧台以及菜单，都被想像成一张超薄的纸被撕开然后再折叠。置身于喷有绿叶图案的高高的吧台、小圆桌以及流畅的绿色墙面贴画中，好像真的可以呼吸到自然的空气。

J.K. Salon

J.K.Salon

地点/广州　面积/150平方米　设计师/林嘉璐
设计单位/力图设计顾问有限公司
主要材料/白机片、亚克力、花纹焗漆板

In the design style, the designer does not impose any kind of "the good reputation" of so-called style, but simply interprets some beautiful things through new materials, introduction of special processing methods, combination of flat organic forms and lights, and rational application of colors and materials.

At the entrance, the ceiling is designed with a deep-color artificially drawn picture of a long-haired girl with wooden effect to highlight the theme of Salon, and set off the main tone of white better in the whole space. The light acrylic wall with white organic patterns, the light reception desk with white curve patterns, the hair beauty platform with white set feet of irregular shapes, the semi-hanging classical carved mirror of white dully polished acrylic, concave-convex baking enamel wallboards of white free curves… all of these like a partita of white melody playing leisurely in the whole space, and it is both modern and elegant.

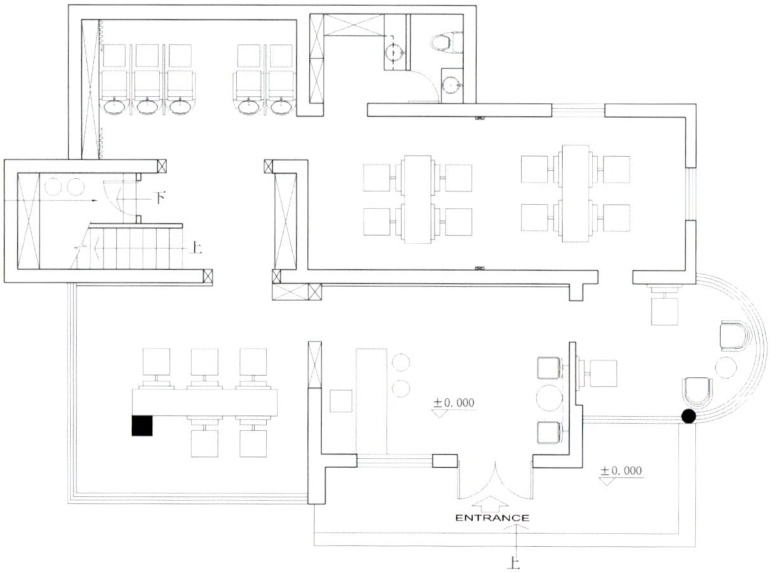
下
上
±0.000
±0.000
ENTRANCE
上

在设计风格上，设计师并没有为其强加任何一种所谓风格的"美名"，只是通过新的材料与特殊的加工方式的运用、平面有机形态与灯光的结合、色彩与材质的合理运用来诠释一些美的事物。

入口天花以一幅深色调的、全人工绘制的、木刻效果的长发少女画来突显Salon之主题，并更好地把整个空间的白色主基调衬托出来。白色的有机图案渗光亚克力形象墙，白色的曲线花纹渗光接待台，白色的带不规则造型台脚的剪发台，半吊着的白色磨砂亚克力古典雕花镜，白色的自由曲线凹凸焗漆墙板……所有的这些，就如同一支白色旋律的变奏曲，在整个空间中悠然跃动，既现代又优雅。

T.Y.M (International) Hair Salon

T.Y.M(国际)发型

The design concept of T.Y.M (International) Hair Salon requires simple, bright and modern style. The main tone is large area of white and small area of black series, collocated with the grass green on the main logo wall to fully show the fresh, fashionable and environmentally-friendly design concept of this hair salon space. The mirrors on the walls and columns not only enhance the sense of extension in this space, but also control the project cost. This project seeks to achieve a special space interpretation of a modern hair beauty center through simple design.

地点/福州　设计师/林洲
设计单位/福州多维装饰工程设计有限公司
主要材料/白色防水镜、白色底细纹仿古砖、镜面板、硅酸钙板、折纸状人造石台面

T.Y.M(国际)发型

T.Y.M(国际)发型的设计定义要求简洁、明亮、现代。主体的色调为大面积的白色和小面积的黑色系，配以主形象墙的草绿色，充分体现了发型空间的清新、时尚、环保的设计理念。墙壁和柱体的镜面不仅增强了空间的延伸感，而且也控制了工程造价。本案力图通过简约的设计来达到对现代发型中心的另类空间诠释。

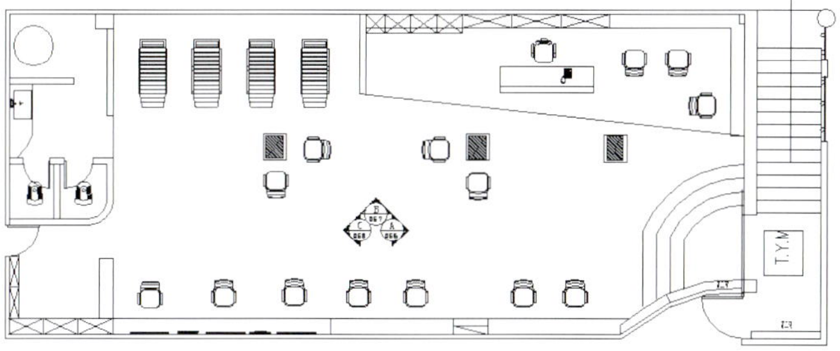
T.Y.M

T.Y.M

Bowling Gym
保龄球馆

地点/苏州　面积/1400平方米　设计师/程通
设计单位/苏州工业园区一通建筑装饰设计有限公司
主要材料/红色马赛克、茶镜、不锈钢、阻燃布拼花硬包、张拉膜灯光变色、软膜天花灯、黑色金属网格、特殊墙纸、特殊荧光地毯、特殊肌理漆、专用地板

This project is the first innovation-model bowling gym in our country. The plane design depends on the three-dimensional space, and the whole space is divided into three functional areas, a hall, a bowling area and a billiards area, which are independent of each other with organic communication with each other. The colors and the shapes interweave each other to endow the space with life and tension.

The color is a bright spot, and the bright red, colorful mosaics and stars-striped carpet contrast each other strongly, which are a paradoxical integration on color and vision.

Light and color are the integration on vision, and the ring-shaped open island-style service desk is dotted with color-changing LED lights. There is a membrane-structure ceiling above the service desk, while built-in LED color-changing lights make the space more artistic. The bowling area is decorated with two lighting designs, one is the normal lighting, and the other is the visual purple lighting. The bowling alleys are designed with special materials which can show their characteristic of lighting in the visual purple lights, and the floor is paved with custom-made fluorescence carpet to cooperate with functions.

In the selection of architectural space, the ceiling and floor, different materials and different colors interweave and echo with each other to create a unique artistic atmosphere.

本案为国内第一家创新模式的保龄球馆。平面构思取决于立体空间。整个空间分为大厅、保龄球区和台球区三个功能区，相互独立而又有机地互通联系。色彩与形体相互穿插，赋予了空间生命与张力。

色彩是此次设计的亮点，鲜艳的红色同斑斓的马赛克和星条状的地毯对比强烈，但却是色彩与视觉上矛盾的统一体。

光与色，是视觉的统一体。环状的开放的岛形服务台上点缀着变色的LED发光二极管。服务台顶上是膜结构的天花，同时内置的LED变色光，使空间更具艺术性。

保龄球区采用了两路照明设计，一路为正常灯光，一路为紫光照明。球道表面采用特殊处理，在紫光下会呈现发光的特点，地毯采用定制荧光地毯配合功能。

建筑空间选型上采用了顶地穿插呼应、材料穿插呼应、色彩穿插呼应，形成了独特的艺术氛围。

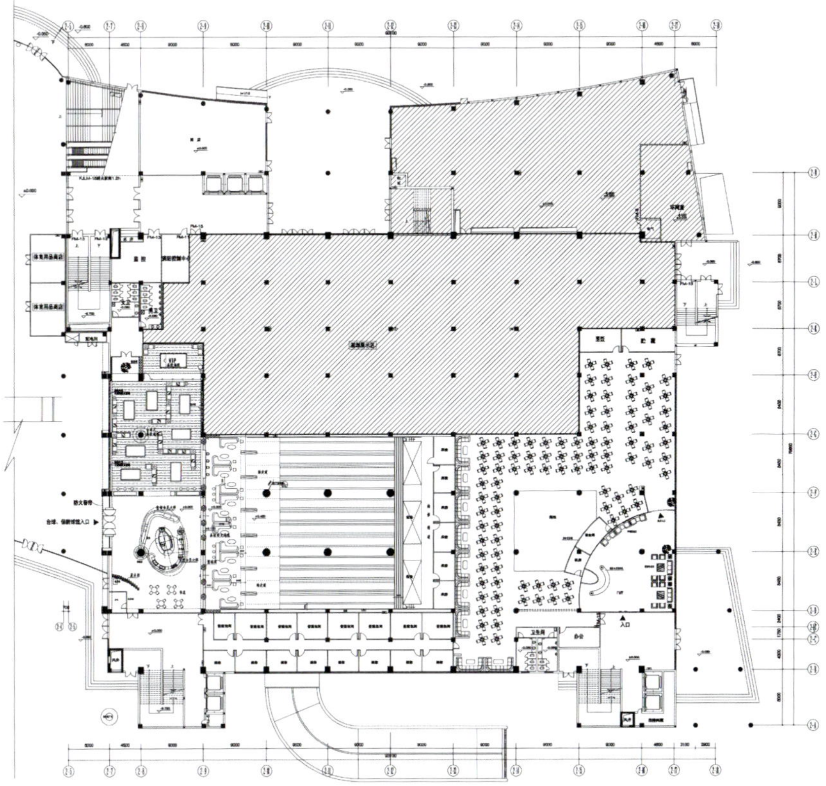

REGAL

REGAL

Fuzhou Dynamic Beauty Salon

福州动感形象沙龙

In this design, the space height is highlighted as far as possible due to the higher height of layer, at the same time showing the simple and clean space background. In order to make this fashion show the noble and elegant flavor further, the red sofa with quite classical charm is introduced into the late soft decoration, so that our guests can enjoy a quiet and relaxed mental feeling which is full of vigor. For the feature of the originally deep architectural plane, the space is designed with a fire-new space viewpoint in the limited space, breaking the sense of oppression of the original interior space and creating a gentle, beautiful space which is full of modern sense. The space is fixed, but the design is full of vitality. Our designer precisely plans out each proportional function space, and a practical and modern space layout presents in front of us.

地点/福州　设计师/金舒扬
设计单位/福建国广一叶建筑装饰设计工程有限公司
主要材料/不锈钢、地砖、大理石

由于有着较高的层高，因而设计时尽量突出空间高度，同时又使得素净的空间背景得以表现。为了让这种时尚更显出典雅高贵的情调，在后期软装饰中采用了颇具古典韵味的红色沙发，让客人有宁谧、轻松而又充满活力的心理感受。原有的建筑平面进深较长，针对这一特点，设计师在有限的空间中，以全新的空间视点进行设计，突破原有室内空间的压抑感，营造出一个柔美、富有现代感的空间。空间是固定的，但设计却是充满活力的。设计师在有限的格局中，精确地规划出每个比例的机能空间。一个实用、现代的空间布局呈现在眼前。

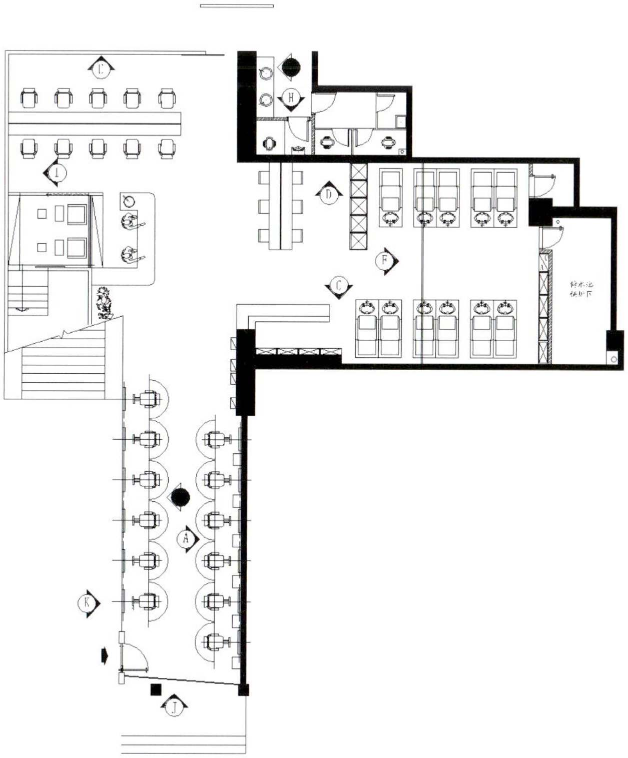

GOLDWELL

Futu Bridal Shop

富图婚纱店

地点/深圳　面积/100平方米　设计师/王锟
设计单位/深圳市艺鼎装饰设计有限公司
主要材料/水曲柳刷白、法国木纹石、灰镜、墙纸、涂料、镜钢
摄影师/钱翔

Love is sacred and pure. A huge amount of white is applied in this space, collocating with faint lighting in the hall. The stretch of white comes into view first, decorated with a small number of trees and leaves, as if it is the legendary Garden of Eden. The design of windows with large area glass is introduced surrounding the overall space, coupling with curtain-like doffing and lighting effects, and it seems like love which is gleaming and elusive, bringing us space experience with unlimited aftertastes. The photos witnessing the beautiful love all the way lead us to the dress show area. In the solemn black, there are Western-style white wedding dresses and Chinese colorful full dresses, and the application of mirrors and steel and the changes of colors are dream-like, both true and fake. The working area is the place for inspiration, and the dream-like decorations of air bubbles rising on the wall highlights the top-created aesthetic personality further.

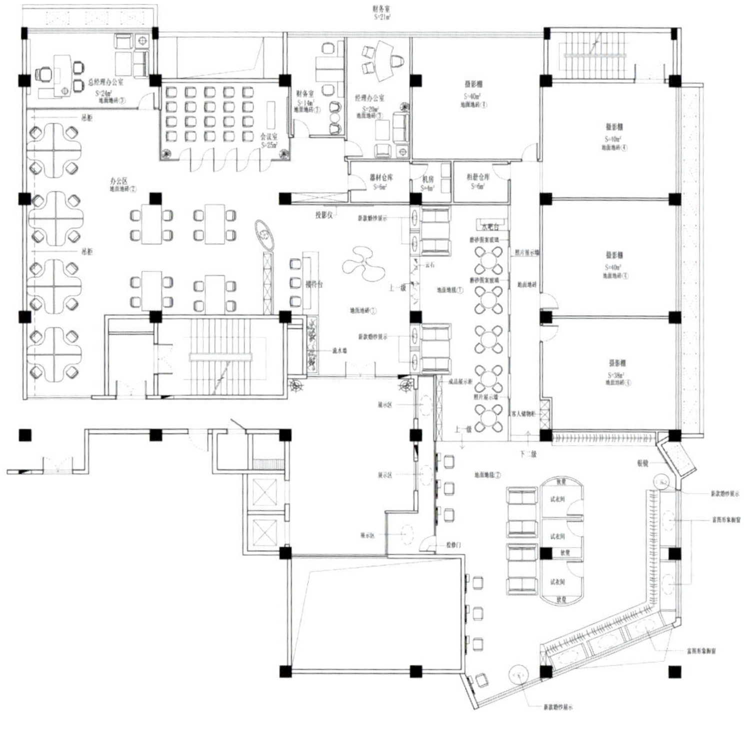
财务室
S=21m²
总经理办公室
S=24m²
财务室
S=14m²
经理办公室
S=20m²
摄影棚
S=40m²
会议室
S=25m²
办公区
器材仓库
S=6m²
机房
S=4m²
相册仓库
S=6m²
投影仪
新款婚纱展示
接待台
上一级
流水墙
成品展示柜
照片展示墙
客人储物柜
展示区
下二级
试衣间
摄影棚
S=38m²

FUTUV·COM

爱情是神圣的，是纯洁的。设计师运用大量的白色，配以大厅内幽幽的灯光。首先跃入眼帘的是绵延的白，加以少量树木、树叶装饰，仿佛就是那个传说中的伊甸园。整体四周采用了大面积玻璃窗设计，加以帷幕般的落纱，配合灯光效果，如同爱情般若隐若现、不可捉摸，给人以无限回味的空间感受。一路见证着美好爱情的相片步入服装展示区。在肃穆的黑色中，既有洁白的西式婚纱，又有色彩斑斓的中式礼服。镜钢的运用，色彩的变幻，如同梦境般亦真亦假。办公区是一个诞生灵感的地方，在墙壁上有如同梦想般上升的气泡装饰，更加突显了高端创意的唯美个性。

FUTUV.com
FUTUV.com

FUTUV.com

FUTUV.com

Gama Photography Gallery

伽玛影研

地点/江阴　面积/300平方米　设计师/陈晓辉
设计单位/品意堂装饰工程有限公司
主要材料/丝绸、木料、大理石

Pure and Romantic Sentiment

The final result of prosperous space is silent crooning only, and it may be just the true realm, just the ultimate form of return to innocence. It gets rid of noise and clutter, which makes the low-key colors to be a thought for imagination and ratiocination, and the good texture and layout are used to induce life into the next station. Choose a color which is red-hot; choose a way of life which is true; choose a design which is thinking-concentrated. The combination of wooden materials and silk is novel without affectation. The color is very quiet, simply quiet, and apart from the soft and light wood color there is no interference. All the materials and all the feelings are heightened by the faint wood color, which is quiet and full of tension. The selection of lights abandons the conventional approach to bring new thought into the space. The quiet and bright space is not only the bouquet waiting right here, but also the memories about youth are infiltrating step by step and become clear gradually. Are the warm yellow color and faint light the superposition of colors or the review of romantic? The pleasant atmosphere makes thinking slow, but our mood is obviously happy.

纯粹而浪漫的情愫

空间繁华到了最后，只剩下内敛的浅吟低唱，这也许才是真境界，才是返璞归真的最终形式。摆脱喧嚣与纷杂，让低调的色彩成为了一个可供想像和推理的思维，借助良好的质感和布局，使生活进入下一个驿站。选择一种色彩，炽热的；选择一种生活方式，真实的；选择一种设计，浓缩思想的。木质材料与丝绸的搭配，新颖而不做作。色彩很静，静得纯粹，除了柔淡的木色之外，没有任何的干扰，所有的材质，所有的感觉，被一抹淡淡的木色渲染，沉静而充满张力。灯的选择，摒弃了常规做法，给空间带来了新的思考。宁静而明亮的空间，不仅仅是守候在旁的花束，关于青春的记忆正一点一滴地渗入，逐渐明晰。暖暖的黄，淡淡的光晕，是色彩的叠加？还是浪漫的重温？愉悦的氛围让思维变得迟钝，但心情却分明是快乐的。

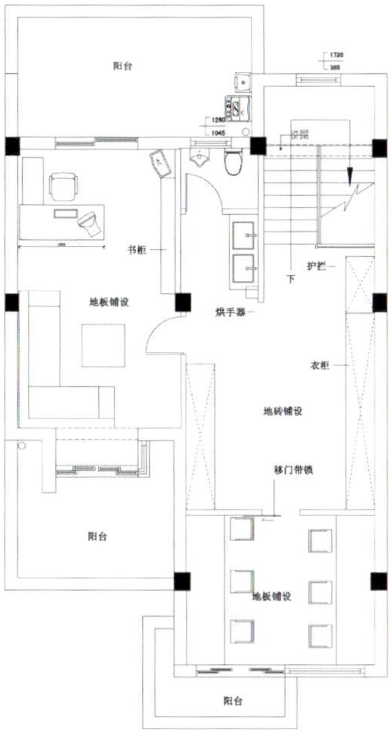
阳台
书柜
地板铺设
烘手器
护栏
下
衣柜
地砖铺设
移门带锁
阳台
地板铺设
阳台

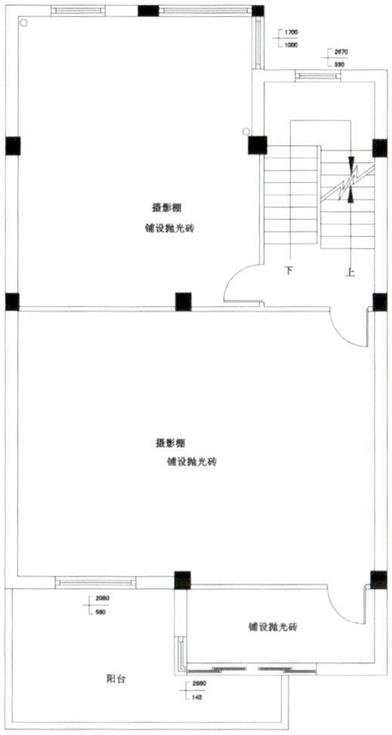
摄影棚
铺设抛光砖
下
上
摄影棚
铺设抛光砖
铺设抛光砖
阳台

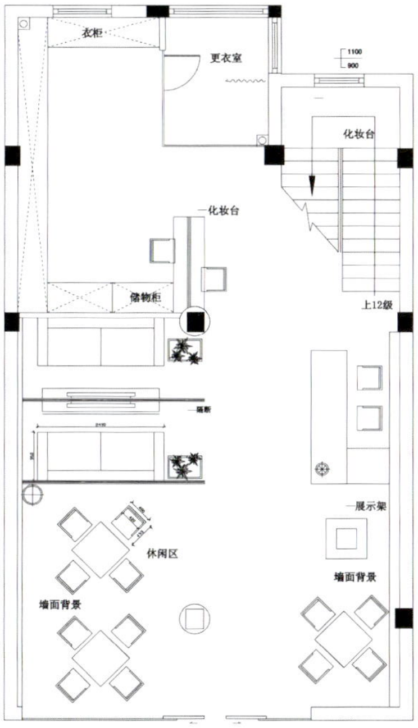
衣柜
更衣室
化妆台
化妆台
储物柜
上12级
展示架
休闲区
墙面背景
墙面背景

YOYI Book

深圳友谊书城

地点/深圳　面积/10000平方米　设计师/陈武
设计单位/深圳市新冶组设计顾问有限公司
主要材料/石膏板、蓝色水性涂料（亚光）、白色水性涂料（亚光）、木板、创作壁纸

Our designer shakes off the monotonous and traditional decorative style for bookshop operation, and on the base of point, line and surface, skillfully integrates the materials (the white aluminium plate and Mongolia black marble) to fully show the quality of this bookshop. The rich secondary source makes the level of the whole appearance more prominent, so that we can feel the inadvertently hidden rhythm and melody. Parts of the interior space are designed with gray, yellow and blue baking-paint glass as decoration on the surface of the columns, and collocating with the assistance of light-colored VI. They collide and echo with each other in the whole interior space, which brings us an easy excitement and pleasure on perception and fully shows the enterprise's brand image and sense of quality.

Glasses Center Of
YAMAHA（雅马哈）专卖

设计师摆脱千篇一律经营书店的传统装修模式，以点、线、面为基础，巧妙结合材质（白色铝板、蒙古黑大理石），完美表现书店品质。丰富的二次光源让整个外观的层次更加突出，使人体会到那些隐藏于不经意间的节奏和韵律。室内部分运用灰、黄、蓝三种颜色的烤漆玻璃作为柱饰面，配以浅色VI的辅助形，在整个室内空间内相互碰撞、相互映衬，在感观上给人一种轻度的兴奋和愉悦，充分展现了企业的品牌形象及品质感。

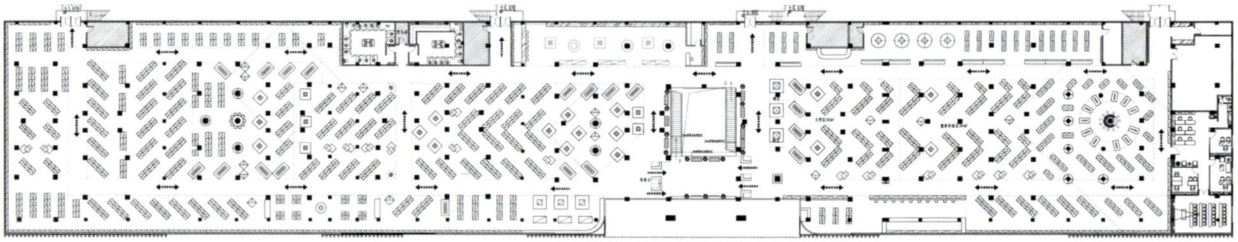

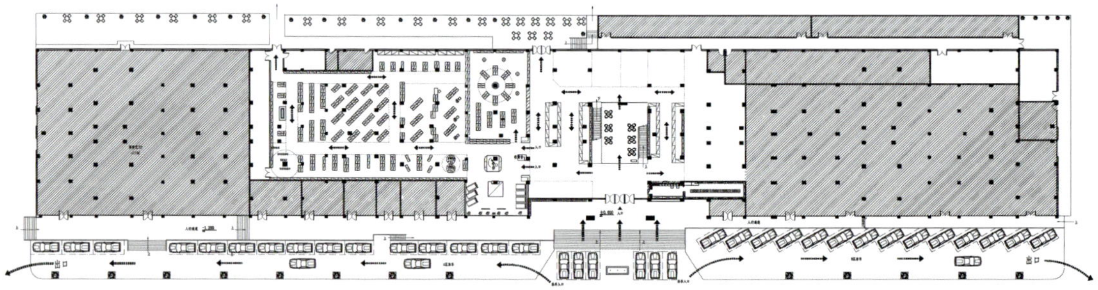

友谊扬帆，启航智慧
友谊书城

初中教辅
初中教辅
初中英语

missionbay coffee
you have choice!

Iridescence o0O

虹色o0O

地点/广州　面积/220平方米　设计师/陈列宝
设计单位/深汕Revolution Studio
主要材料/水泥地坪漆、氟碳涂层、红色夹胶镜、光纤

This project is a hair salon oriented to the young and fashionable consumer group. In the plane design, three red ovals are cut and combined to form the LOGO of the o0O Hair Salon. Part of the interior design is based on red and oval elements throughout the design, from the space mirrors on the wall, to the light-free patterns on the ground, and to the ceiling. Our designer strives to ensure the basic lighting in the work area with minimal lights to create the environment and atmosphere. On the first and second floors, the hollowed-out oval mirrors cooperate with low-wattage droplights with different angles to provide lighting on the first and second floors. In the VIP area on the fourth floor, the work area and walkway are distinguished by fiber, and the walkway is not designed with any lights. The seats are equipped with automatic detection system, so the group of work lights opens once customers sit down on the seat, and self-closes a short while later after the customers' leaving. Our designer hopes that an impressive space can remain in the hearts of customers with future modern style, unified and harmonious collocation of colors, and healthy and energy-efficient interior environment.

本案是一个定位于年轻时尚的顾客群体的发型沙龙。平面设计上，以三个红色的椭圆形进行剪切组合，构成O3 Hairsalon的LOGO。室内设计部分以红色与椭圆为主要元素，从墙面的太空镜造型，到地面的透光图案，直至天花吊顶，贯彻于设计始终。设计师力求在保证工作区足够照明的基础上采用最少的灯光来营造环境气氛。一、二层中空椭圆镜造型，以不同角度的镜面配合低瓦数吊灯，给一、二层提供照明。四层VIP区用光纤来区分工作区和走道，走道完全不设置灯光。座位边设置自动感应系统，顾客坐下即打开工作灯组，离开片刻自动关闭。设计师希望通过未来现代的风格、统一和谐的色彩搭配、健康节能的室内环境，将一个印象深刻的空间留在顾客心中。

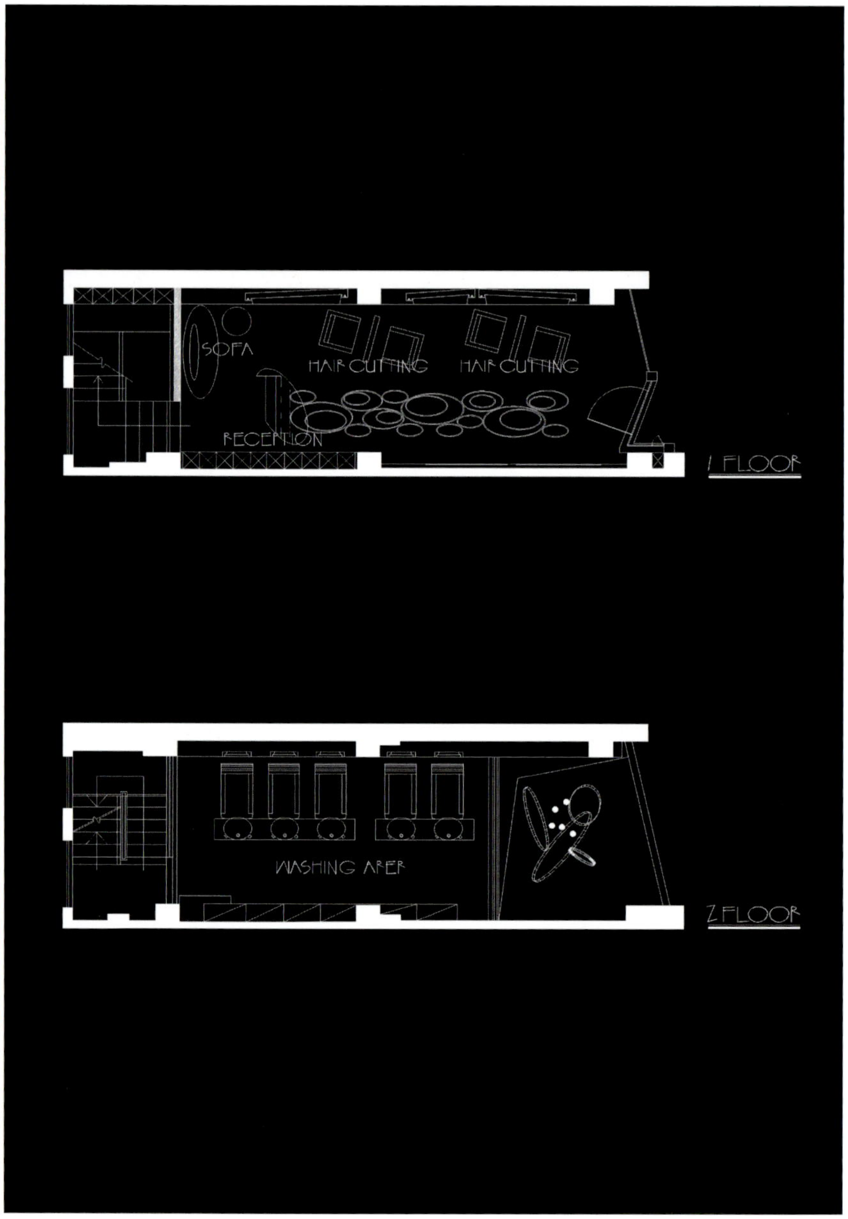
SOFA
HAIR CUTTING
HAIR CUTTING
RECEPTION
1 FLOOR
WASHING ARER
2 FLOOR

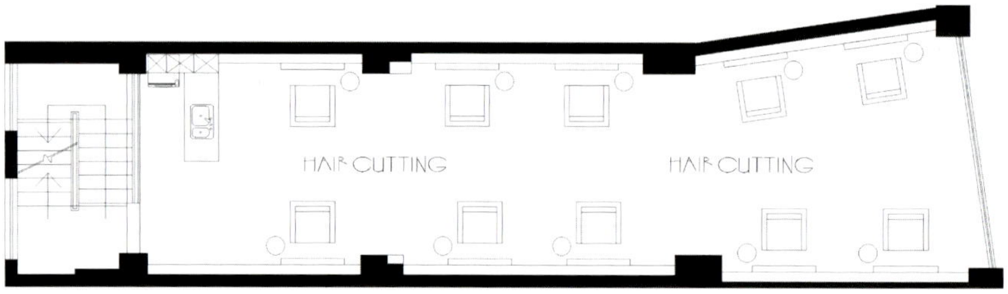

3 FLOOR

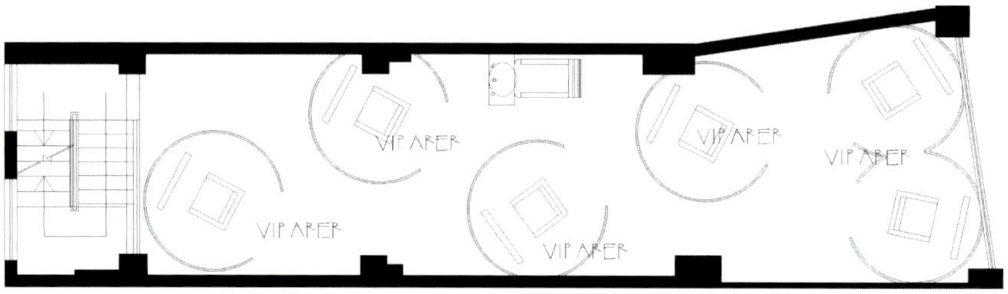

4 FLOOR

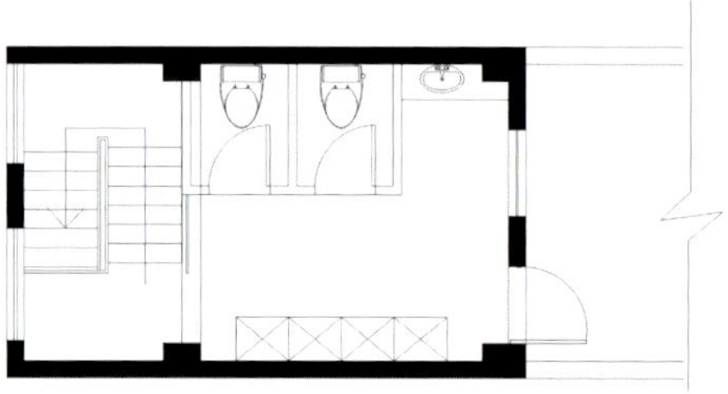

5 FLOOR

List of Popular Jewelry

珠宝流行榜

List of Popular Jewelry in China is the largest diamond wholesale shop in Jewelry Trading Centre in Shenzhen, and in the design concept, the popularity presentation is chosen as the theme to fulfill the largest volume of release and achieve a luxurious and comfortable shopping space. The breakthrough of design is not the beautiful inside and outside image, but the layout of the counter that promotes sales. The target group is the retail customers from different cities nationwide, so the design is made with the avoidance of excessive positioning, and the design with a target is introduced to achieve the purpose of sales.

地点/深圳　面积/160平方米　设计师/艾永庆、于蕴昊
设计单位/深圳市荣信商业空间设计公司
主要材料/大理石、地毯、铜色不锈钢、石膏板、乳胶漆

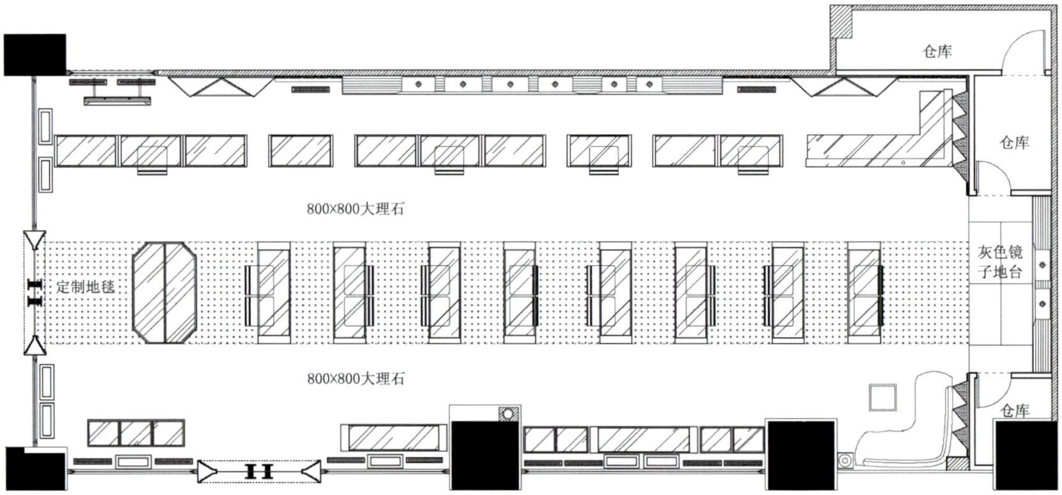
仓库
仓库
800×800大理石
灰色镜子地台
定制地毯
800×800大理石
仓库

中国珠宝流行榜

LIST OF POPULAR J

詩樂園
DIAMOND PARK
7－8月

中国珠宝流行榜是深圳珠宝交易中心最大的钻石批发店，设计概念以流行发布会为主题，满足最大的放货量，实现奢华、舒适的购物空间。设计的突破，并非是华丽的内外部的形象，而是这个能够促进销售的柜台布局。销售人群针对的是全国不同城市的零售客户，设计回避过高定位，用针对性的设计达到销售目的。

出口
EXIT
7-8月
LIST OF POPULAR JEWELRY IN CHINA

DIAMOND PARK
7-8月
中国珠宝流行榜
LIST OF POPULAR JEWELRY IN CHINA
LIST OF POPULAR JEWELRY

7–8月

中国珠宝流行榜
LIST OF POPULAR JEWELRY IN CHINA
7–8月
LIST OF POPULAR JEWELRY IN CHINA

B&F Cosmetics Franchised Store in Zhongxin Square

中信广场爱敬化妆品专卖店

It is a franchised store of the Korean BM Bellecia cosmetics brand, and in the application of colors, black is used to highlight the products, while golden is introduced to enhance the high quality of products. The distinguished golden statue is embedded in solemn and mysterious mirrors, and golden is used to set off the luxury of high-grade cosmetics and bring us gorgeous sense of magnificence. On the simple and agile show cabinets and platforms which are mainly designed with black geometric figures, white lights shine on the products, and the contrast of black and white highlights the delicacy and meticulousness of the products.

地点/广州　面积/40平方米　设计师/龙鹏飞
设计单位/概念空间设计工作室
主要材料/黑玻璃、玻璃钢、浮雕、地板砖、难燃板、软包、灯箱

这是韩国 BM Bellecia化妆品品牌专卖店。在色彩的运用上，运用黑色是为了突显产品，而金色是为了提升产品的贵气。尊贵的金黄塑像镶嵌在庄严神秘的黑镜中，以金色衬托出高端化妆品的奢华，给人以富丽堂皇的华丽感。在以黑色几何图形为主的简单利索的展柜、展台中，洁白的灯光照在产品上，黑白的强烈对比突出了产品的精巧与细腻。

BM Bellecia

BELLECIA
Facial Essence

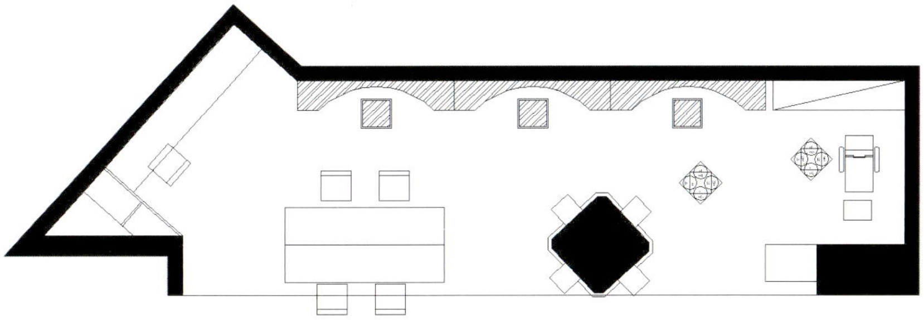

Japanese WATABE (Shanghai) Wedding Studio

日本华德培(上海)婚庆工作室

地点/上海　面积/330平方米　设计师/吕永中
设计单位/唯品设计
主要材料/杜邦特殊纸张、境面、不锈钢、木材、人造石、布艺
摄影师/吴永长

This project is designed to create a fire-new space for experience of the personalized wedding. The interior space plan applies a comparatively free space layout, and the methods of organizing space such as enclosing, dissociating, and separating and so on are used to integrate different functional areas. The center part of the first floor is the reception hall covering nearly the half of the area, and the "paper bounding wall" of custom-made DuPont Specialty Paper is used to enclose each area. The wall with the layout of oval-shaped surface controls the space effect of the whole floor, and it is the visual image of the whole space. The arc-shaped wall is disconnected to form an opening without any border, and extends to the other two areas. The initial starting point of paper wall is the textile structure of wedding lace, and its translucent effect is formed by the relations of textile density, which is soft and rich in texture. In the space of three floors, the shapes and materials are arranged according to the functional layout to provide a wide range of creative backgrounds to choose for wedding interior photography. The egg-shaped reception desk implies the naissance and consummation of marriage, as well as brings a little gentleness and romance into the comparatively foursquare of the three floors, and the sleek and smooth surface echoes with the rich texture of the paper wall.

本案欲创造一个崭新的个性化创意婚庆体验空间。室内空间规划选择了较自由的空间布局，用围合、游离、分离、相望等空间组织手段，把不同的功能区域加以整合。一层中心部位是接待大厅，占据了近一半的面积，空间围合是采用定制的杜邦特种纸做成的"纸围城"，平面呈椭圆形状布局，控制着整个一层的空间效果，成为整个空间的视觉形象。圆弧形城墙断开处成为开口，没有边界，并延展到另外两个区域。纸墙设计的最初出发点是婚纱蕾丝的纺织结构，其半透明状的结果是由纺织的疏密关系形成，质感丰富而轻柔。三层空间根据功能布局去安排造型及材质，为婚纱室内摄影提供了多种创意背景的选择。蛋形接待台暗示婚姻的诞生与圆满，同时为三层较方正的空间带来一丝柔性与浪漫，光洁圆滑的表面与纸墙丰富的肌理形成对比。

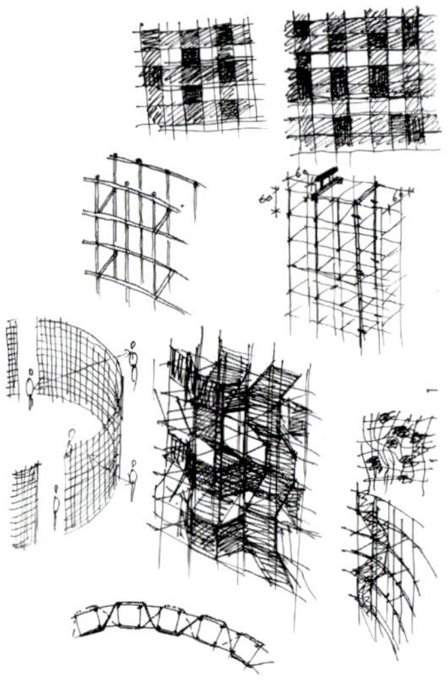

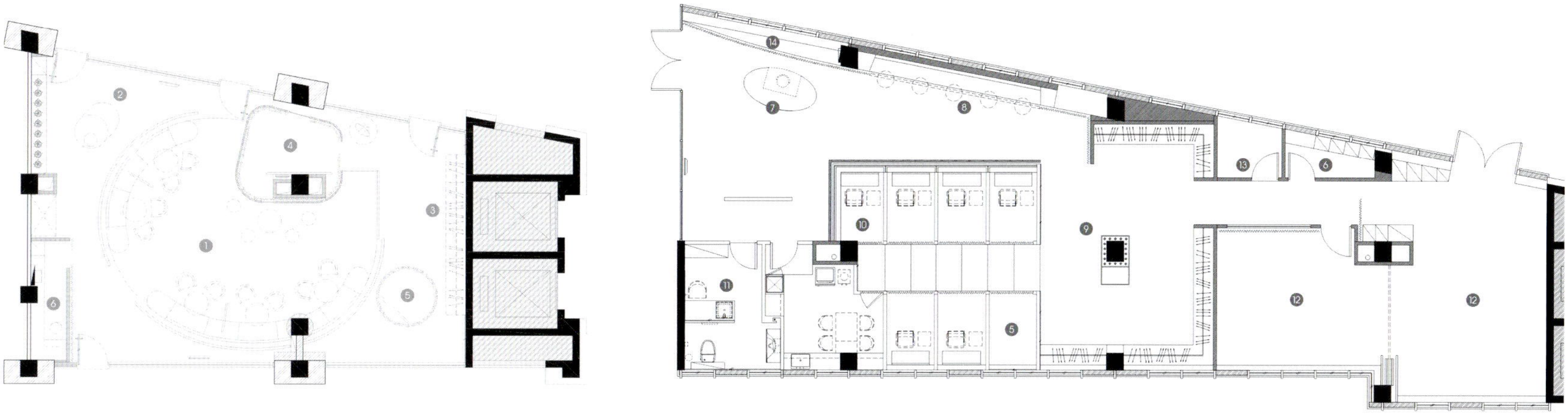

❶ 接待大厅
❷ 花店
❸ 婚纱展示区
❹ VIP房
❺ 试衣间
❻ 工作间
❼ 前台
❽ 书吧
❾ 婚礼服装区
❿ 化妆间
⓫ WC
⓬ 摄影棚
⓭ IT
⓮ 储藏区

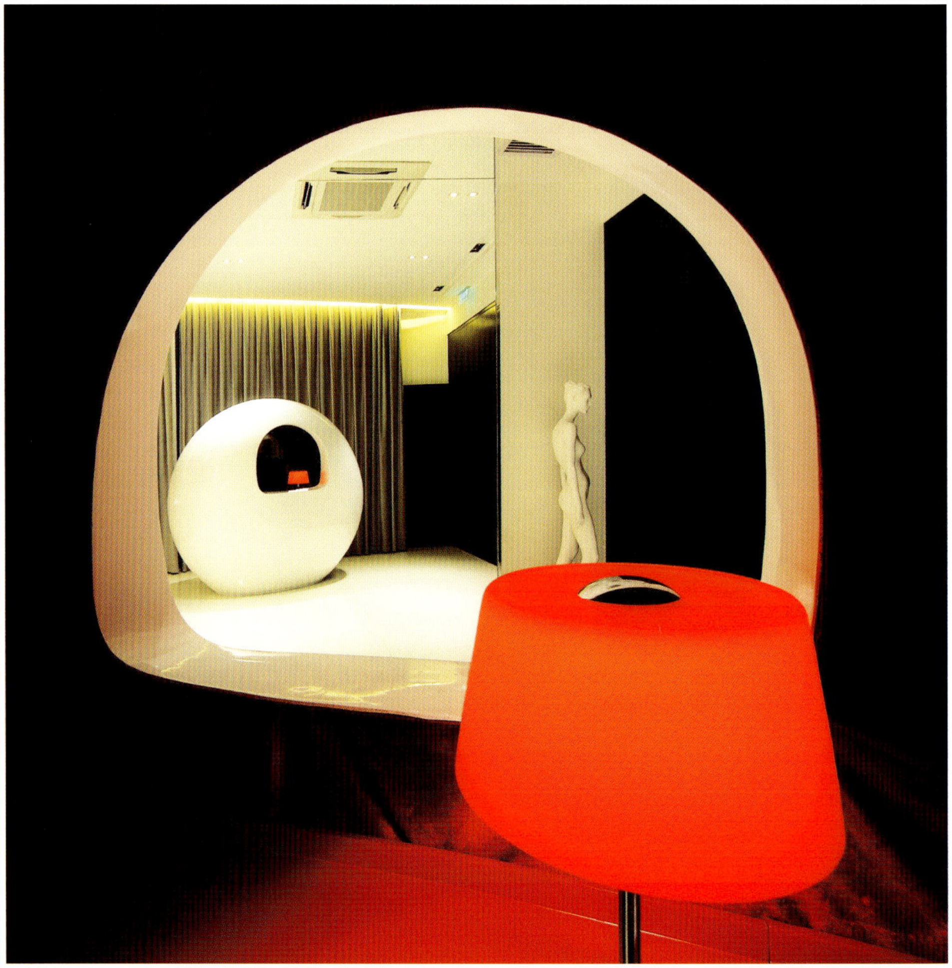

Pure Pearl Store

千足珍珠示范店

地点/诸暨　面积/300平方米　设计师/龚小刚
设计单位/龚小刚设计师事务所
主要材料/雪花白大理石、中式纹样木隔扇、草编墙纸、中国人物工笔画、金卤灯、LED灯、中式荷花灯、黑色线绳、纱帘

Simplicity, delicacy and elegance are the first impressions of Pure Pearl Store. Pearl is the point, the bead chain is the line, the bead curtain is the surface, and the light, curtain and decorative wall are designed with shapes of points, lines and surfaces which are correlative to the pearl. "The lotus is clean even growing in the muddy pond. It is so pure, delicate and bright. The lotus is consistent, continuous and coherent deep inside. It appears to be straight, proper and honest. It gives a fantastically good smell and people could even sense its excellent smell far away. It has no unnecessary branches. It can be only appreciated distantly but not touched blasphemously." The lotus symbolizes the noble and pure, and the lotus droplight, lotus decorative wall and the pastel green light from the show cabinet make the light white fluorescence from the pearl, green lotus seedpods and pink petals echo with each other, and make the beauty of the pearl more conspicuous, holy, dignified and elegant. Lighting design is another bright spot of design in it, light purple, light blue, pastel green... multilevel lights are designed to show the delicate and precious pearls.

PURE PEARL

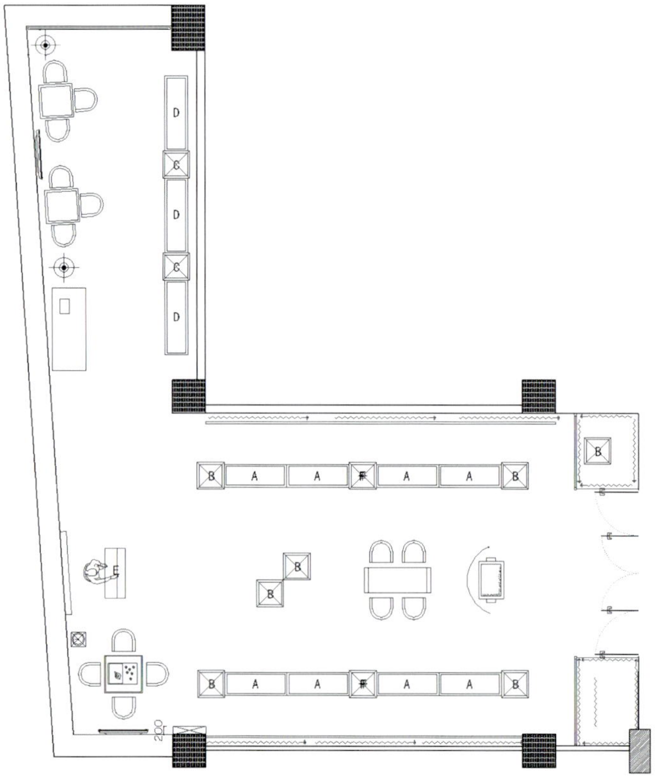
D
C
D
C
D
B
A
A
A
A
B
B
B
B
E
B
A
A
A
A
B

EARL

PUREPEARL

简约、细腻、古雅是千足珍珠示范店给人的第一印象。珍珠是点、珠链是线、珠帘是面，设计师将光、帘与饰面墙设计成点、线、面这些与珍珠相关的形状。“出淤泥而不染，濯清涟而不妖，中通外直，不蔓不枝，香远益清，亭亭净植，可远观而不可亵玩焉”，荷花象征着高洁，荷花吊灯、荷花装饰墙、展柜发出的粉绿色的光，使珍珠放射出来的淡白色荧光与绿色的莲蓬、粉嫩的花瓣相互辉映，使珍珠的美丽更为出众、圣洁、端庄、雅致。灯光设计也是本案的一大设计亮点，淡紫、淡蓝、粉绿……多层次的灯光都是为了体现珍珠的细腻与珍贵。

Croissants de France in Vanke Town

可颂坊万科城店

This project is based on wine red sign to highlight the colors and lead guests' line of sight. The display counter is decorated with strong lighting to become a stage in the store, and the large murals of French flavor own strong visual effects and emphasize the brand from France. The walls are designed with beige color to enhance the sense of texture, and the spotlight sources are used to make the space suitable for negotiation, business and leisure activities. The whole decoration makes our guests fascinated by the elegant decoration and comfortable environment at the entrance. In this project, our designer puts forward "concept of the third space", that is to strive to design this store to be a space between work place and living place, so that it is a space suitable for negotiation and business activities, relaxation, and abandoning all the troubles temporarily, which creates a space with fashionable, elegant and diversified experience of French cultural background.

地点/深圳　面积/120平方米　设计师/黄群生
设计单位/深圳市华极室内设计有限公司
主要材料/仿古砖、酒红色防火板、粗面肌理涂料、胡桃木

谢 谢 惠 顾
Thank you for patronizing CDF
取阅后请放回

现烤现卖
可颂坊
Croissants de France
新上市
New Taste

Thank you for patronizing CDF

谢 谢 惠 顾
Thank you for patronizing CDF

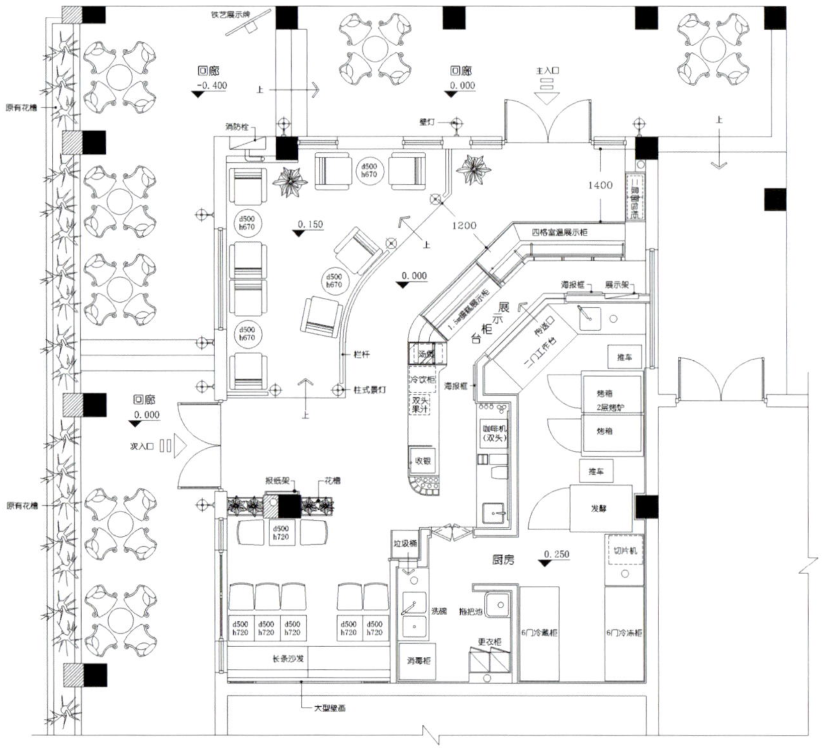

回廊
主入口
次入口
厨房
长条沙发
大型壁画
原有花槽

Croissants de France

Croissants de France

Croissants de France
Croissants de France

MONTPARNASSE
CALAMITY JANE

MONTPARNASSE
CALAMITY JANE

本案以酒红色为主调招牌，强调色彩，吸引客人视线。展示柜台配以强烈灯光，使其成为店中舞台。大幅法式风情壁画，有强烈视觉效果，并强调品牌源于法国。墙身采用米白色，增加肌理感，并使用聚光灯源让空间适合交谈、商务、休闲活动。总体装饰使客人一进入就能被典雅的装饰和闲适的环境所吸引。在本案中，设计师提出“第三空间的概念”，就是努力把本店做成工作场所与生活场所中间的空间场所，使其成为适合交谈和商务活动、放松心情、暂时抛开一切烦恼的空间，营造出有法国文化背景、时尚、典雅、多元化的体验空间。

Croissants de France
Croissants de France

可颂坊
Croissants de France

Mingcheng Tea Supermarket

茗城茶业超市

地点/汕头　面积/3800平方米　设计师/蔡烈波、张育莲
设计单位/伊诺设计装饰有限公司
主要材料/柚木板、实木通花、墙纸、天然竹片、仿古砖、抛光砖、黄锈石板、浅啡网石板、火烧石板、火烧麻石、茶色玻璃

Drinking tea can make us mentally and spiritually calm and help to cultivate our character and remove distracting thoughts, which is in step with oriental philosophical thinking that advocates quietness, and it is also in line with "spiritual introspection" ideology of Buddhism, Taoism, Confucianism. Drinking tea is to satisfied one's thirst with tea as drinks. Tasting tea pays attention to the color, fragrance and flavor and is particular about the quality of water and tea set, and we can taste carefully when drinking. Tea culture is particular about the environment, atmosphere, and music, brewing skills and interpersonal relationships and so on. The highest level—the tea ceremony, the philosophy, ethics and morality are integrated into the activities of drinking tea, and we cultivate our moral character and nature and taste life through tasting tea, and then achieve the spiritual enjoyment. In Mingcheng Tea Supermarket, the atmosphere of antique beauty, fresh and natural layout, multilevel lighting, antique patterns, the stocky show cabinet, tea seats and chairs, all of which show the spirit of tea ceremony and propagate the oriental culture.

抢新鲜
FRESHNESS TEA
新茶上市

喝茶能静心、静神，有助于陶冶情操、去除杂念，这与提倡清静的东方哲学思想很合拍，也符合佛道儒的“内省修行”思想。喝茶：将茶当饮料解渴。品茶：注重茶的色香味，讲究水质、茶具，喝的时候又能细细品味。茶艺：讲究环境、气氛、音乐、冲泡技巧及人际关系等。最高境界——茶道：在茶事活动中融入哲理、伦理、道德，通过品茗来修身养性、品味人生，达到精神上的享受。茗城茶业超市古色古香的氛围，自然清新的布局，多层次的灯光，以及古老的纹样，敦实的展柜、茶座、茶椅都力图体现茶道的精神，弘扬东方文化。

东方名优茶

MING CHENG TEA MALL
新茶上市

蘭茶坊
蘭茶坊
特价品
贈送品

順慶茶莊

CHENG
FRESHNES
38

Dongpeng Sanitary Appliance

东鹏洁具

地点/佛山　面积/800平方米　设计师/周晓鸥
参与设计/易兆祥、陈钰基、莫荣球、张小虎
设计单位/佛山市墨象设计顾问有限公司
主要材料/马赛克、东鹏白洞石、大理石、柚木
摄影师/杜锐华

Water has invaded into human thoughts and culture with connotation of its original essence, and in the long history, water has been sublimated into a spiritual state along with evolution. We can catch the humanity from the hydrology, where water area must be the place of economic and cultural prosperity, and it is water that makes our Chinese nation develop and multiply to endless generations. Once own a place, you can achieve a great career lasting for thousands of years. Dongpeng Sanitary Appliance Headquarter is composed of "Waterscape · Jade Pool with Treasures", "Waterscape · Hustle Flow", "Water Waving with Wind Blowing" and "Gold Dew from the Sky", and the harmonious and compact layout of space and careful and meticulous design and decoration integrate the neo-orientalism culture and Bali South Asian culture into a whole, which brings us fresh and rich artistic atmosphere; the waterscapes throughout the design echo with each other and self-unitize separately, which fully interpret "water culture" initiated by Dongpeng Sanitary Appliance. The rushing, floating or ding-dong sound of water add little vividness and fun simply into the gentle space.

东鹏洁具
Dongpeng

水以其原始的精髓内涵已渗入人类文化思想中，在漫漫的历史长河中，伴随着进化，水已经升华为一种精神境界。由水文看人文，水流域必是经济、文化繁荣之处，是水使中华民族得以发展繁衍，世代不息。能拥一方水土，必成一方伟业，千秋万载。东鹏洁具总部由“水景·聚宝玉池”、“水景·川流不息”、“风生水起”、“天降金露”构成，以和谐紧凑的空间布局，周密细致的装饰设计，将新东方主义文化与巴厘岛南亚文化融成一体，带来了清新浓郁的艺术氛围；与贯穿始终的水景设计相互呼应，自成一体，充分演绎了东鹏洁具所倡导的“水文化”。或奔涌或潺缓或叮咚的流水声为舒缓的空间平添了几分灵动与写意。

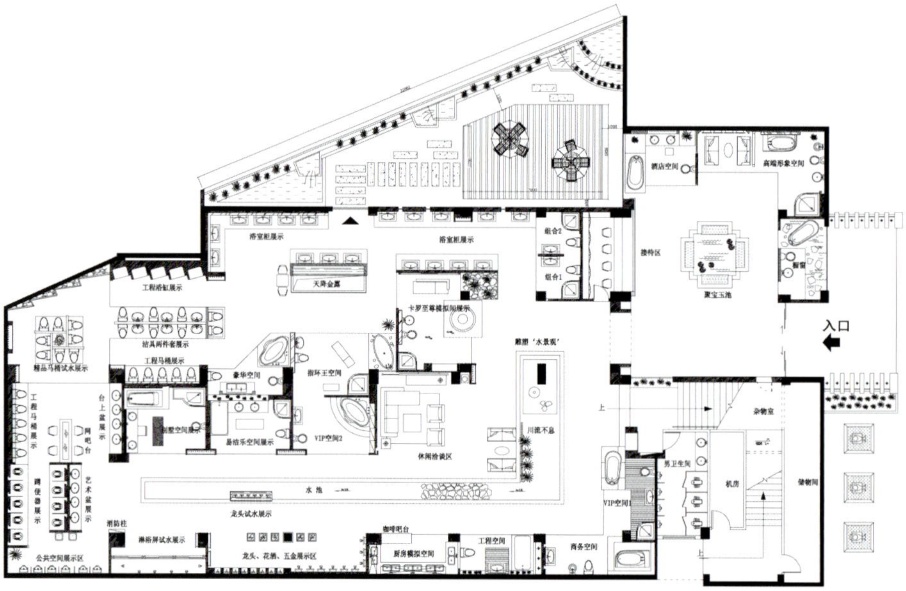

Shandong Jinluo Group

山东金锣集团

Colorful Rings

The design of colorful rings is introduced boldly into this project, and it is collocated with high-tech materials, so that its soft, floating and clean feeling comes to our mind. Under the blue lights, the ring-shaped exhibition stands fully show the beauty of their lines with up-and-down extension and free move, the lines in the air are soft and full of strong texture. The water drop ring-shaped moving style is in line with the beautiful lotus pond on the display wall, where the green lake and fresh lotus flowers are lively and vivid, and the concept of green products enjoys popular support. The top of the recessed ring-shaped design is decorated with a white oval light box to increase the brightness. The artificially operated video equipments are hanging on the black concavity which are divided by a blue light strip to highlight the aesthetics of design. A set of seats are specially designed around the ring-shaped video, so that our visitors can enjoy the corporate culture of our company while relaxing.

地点/山东　面积/4500平方米
设计单位/北京兴中创国际展示有限公司
主要材料/钢材、木料、大芯板、铝型材、铝塑板、免漆木板、喷漆、亚克力板、写真喷绘

污水处理设备
WASTE WATER TREATMENT EQUIPMENT
万利达

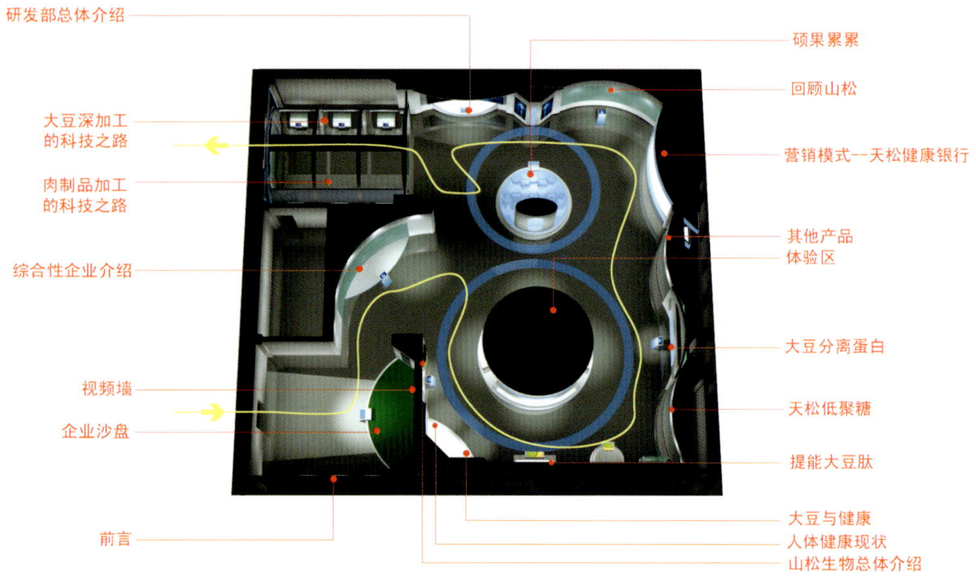
研发部总体介绍
硕果累累
回顾山松
大豆深加工
的科技之路
营销模式--天松健康银行
肉制品加工
的科技之路
其他产品
体验区
综合性企业介绍
大豆分离蛋白
视频墙
天松低聚糖
企业沙盘
提能大豆肽
前言
大豆与健康
人体健康现状
山松生物总体介绍

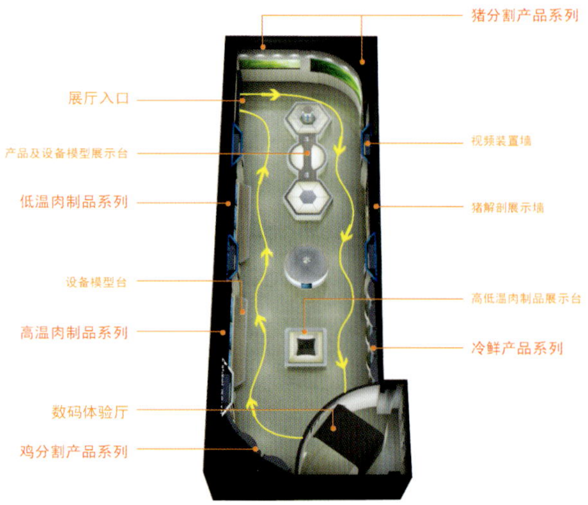
猪分割产品系列
展厅入口
视频装置墙
产品及设备模型展示台
低温肉制品系列
猪解剖展示墙
设备模型台
高低温肉制品展示台
高温肉制品系列
冷鲜产品系列
数码体验厅
鸡分割产品系列

净水处理系统

COMBINED THERAPY

提高人体免疫力

变幻圆环

此展厅大胆尝试运用了变换多样的圆环状设计，搭配高科技材料相映衬，其柔和、流动、洁净之感跃上心头。在蓝色灯光的映照下，环形展台充分展示着线条的美感，上下延伸流动自如，空中的线条柔和中充满着刚强的质感。水滴环形的流动造型恰与展墙池塘荷花美景一致，碧绿的湖水，鲜艳的荷花，生动形象，绿色产品的观念深入人心。凹进式的环形设计，上方各用白色椭圆灯箱以增强其亮度。黑色的凹面中悬挂可人为操纵的视频设备，凹面之间采用蓝色光带加以隔离，突出设计的美感。在环形视频的四周人性化地安排了一排座位，可以让参观者在休息的同时欣赏公司的企业文化。

金锣鸡分割

金锣肉制品
绿色、安全、营养、健康
JINLUO MEAT PRODUCTS

DVD VIDEO

净水处理系统

水处理设备
TREATMENT EQUIPMENT

净水处理系统